NAVIGATING TECHNOMEDIA

New Social Formations
Series Editor:
Charles Lemert, Wesleyan University

Ethnicity: Racism, Class, and Culture
by Steve Fenton

Public Sociology: From Social Facts to Literary Acts, 2nd ed.
by Ben Agger

Shades of Loneliness: Pathologies of a Technological Society
by Richard Stivers

Community on Land: Community, Ecology, and the Public Interest
by Janel M. Curry and Steven McGuire

From Panthers to Promise Keepers: Rethinking the Men's Movement
by Judith Newton

Social Theory and Aging
by Jason L. Powell

Navigating Technomedia: Caught in the Web
by Sam Han

NAVIGATING TECHNOMEDIA

Caught in the Web

Sam Han

ROWMAN & LITTLEFIELD PUBLISHERS, INC.
Lanham • Boulder • New York • Toronto • Plymouth, UK

ROWMAN & LITTLEFIELD PUBLISHERS, INC.

Published in the United States of America
by Rowman & Littlefield Publishers, Inc.
A wholly owned subsidiary of The Rowman & Littlefield Publishing Group, Inc.
4501 Forbes Boulevard, Suite 200, Lanham, Maryland 20706
www.rowmanlittlefield.com

Estover Road
Plymouth PL6 7PY
United Kingdom

British Library Cataloguing in Publication Information Available

Library of Congress Cataloging-in-Publication Data

Han, Sam, 1984–
Navigating technomedia : caught in the web / Sam Han.
p. cm. — (New social formations)
Includes bibliographical references and index.
ISBN-13: 978-0-7425-6023-9 (cloth : alk. paper)
ISBN-10: 0-7425-6023-6 (cloth : alk. paper)
ISBN-13: 978-0-7425-6024-6 (pbk. : alk. paper)
ISBN-10: 0-7425-6024-4 (pbk. : alk. paper)
1. Information technology—Social aspects. 2. Digital media—Social aspects. 3. Virtual reality—Social aspects. 4. Internet—Social aspects. 5. Technology and civilization. I. Title.
HM851.H345 2008
303.48'33—dc22

2007028037

Printed in the United States of America

∞™ The paper used in this publication meets the minimum requirements of American National Standard for Information Sciences—Permanence of Paper for Printed Library Materials, ANSI/NISO Z39.48-1992.

Contents

Acknowledgments

The process of writing (*and* thinking) necessarily includes many, not solely the author credited on the cover. It is best described by Gilles Deleuze and Felix Guattari in *A Thousand Plateaus*:

> Since each of us was several, there was already quite a crowd. Here we have made use of everything that came within range, what was closest as well as farthest away . . . We have been aided, inspired, multiplied.[1]

I would like to acknowledge a few of those who were within range (both close and far away), who, in turn, have aided, inspired, and quite certainly "multiplied" me, in that they have all played an enormous role in shaping my thinking and the writing of this book.

Charles Lemert has been a true mentor and friend. More than anyone, he has influenced my intellectual development. I'm honored to have learned social theory from him; he is surely one of the best. In the writing of this book, he has guided me through it all, from the early sketches to the final edits, always encouraging and supporting me fully. In addition, his wife Geri and daughter Annie have made me feel a part of their family. I could say much more about Charles, but his own work and the numerous projects that he helps to put out speak for themselves. Without him, none of this could even be thinkable.

Besides Charles, numerous people have read every word of this book in various (some not-so-flattering) stages. Early on it was Jonathan Cutler, with whom I traded ideas that are now found in chapter 5, and Anthony Elliott, who provided excellent feedback on the very first version of the manuscript. Later on, Daniel Chaffee and Dan Reif provided very helpful comments and worked hard to find a place to scan an edited manuscript well over a hundred pages into PDF all the way in Adelaide, Australia.

I must also recognize members past and present of the social theory tutorial at Wesleyan University, with whom I shared numerous conversations that found their way into this book. Of them, Makenna Goodman, Eric Hsu, Ben Goldstein, and David Stein deserve particular mention.

At the Graduate Center of the City University of New York, I have the pleasure of studying with Patricia Ticineto Clough, who inspires me to think beyond the confines of sociology and engage in new thought. Her intellectual breadth and rigor (not to mention her utmost support of students) are matched by few. Stanley Aronowitz is a wonderful teacher and ally who is the model of an engaged intellectual. I must also thank my fellow students at the Graduate Center, especially members of the blog Dossier on Method (www.dossieronmethod.com), for countless discussions and for being my intellectual buddies, my fellow theory-heads. The kind of camaraderie we share is what the academy, in general, is sorely missing.

I also wish to acknowledge my colleagues at the Department of Sociology, Anthropology, and Social Work at the College of Staten Island/CUNY, most especially David Goode (for taking a chance on a young academic) and Nancy Vega. It is my students there, however, who make teaching a wonderfully rewarding experience. Their honesty and excitement ensure that it is never boring, but always exhilarating.

At Rowman & Littlefield, I thank Alan McClare for believing in this project from the beginning and Douglas Piccinnini for answering all of my neurotic e-mails with quickness and grace. They and others have made the publication of this book comfortable beyond measure.

Closer to home, I wish to thank my parents Dohei and Michelle Han. Though they cannot read what is written here (with much comprehension at least), I hope that seeing their son's name in print has made decades of back-breaking work in the family dry cleaners worthwhile. My brother Paul has been a constant source of friendship, knowing when to chide and when to encourage.

And finally, being able to write, work, and live with someone like Khalia Frazier is a blessing. Not a day goes by that I'm not truly grateful for her love.

Note

1. Gilles Deleuze and Felix Guattari, *A Thousand Plateaus: Capitalism and Schizophrenia.* (Minneapolis: University of Minnesota Press, 1987), 3.

Series Editor's Foreword

Resetting the Temperature: Are Technomedia Cool or Hot or Other?

Charles Lemert

More than four decades ago, in 1964, Marshall McLuhan's *Understanding Media* took the literary public by storm—a very uncool thing in itself. The book bears reading and rereading even now as the very idea of media has itself been transposed beyond McLuhan's remarkable distinction between hot and cool.

Hot media, he said, are high definition in that they transmit a great deal of information along a single dimension. Thus, radio, the movies, and the photograph are hot. By contrast, television, the telephone, and the cartoon are cool. Cool media are relatively low in information, which is to say low definition. "High definition is the state of being filled with information." What is the difference? Cool media must be filled in because relatively little information is given; hot media supply so much information that they permit very little participation. Cool encourages the imagination; hot dulls it.

Today it can be difficult to follow McLuhan's distinctions because media contents have changed so weirdly. Thus, in 1964 radio was mostly music like Elvis and the NBC Symphony Orchestra, news, and dramatic comedy. There was no talk radio. As far as music was concerned, hip-hop was still searching for Gil Scott-Heron. Television, then, was variety shows, wrestling, JFK's funeral and the like; today it is disinformation dressed up as news, sitcoms, and crime scenes. The changes are not total to be sure. In any case, McLuhan's point, as he famously said, was that the medium itself is the message, with very powerful effects on the human imagination.

Now that most televisual media are digital, most media of the visual kind are the lowest common denominator of definition—a false universe of zeros and ones unfathomably organized somewhere beyond anything we can imagine; where also the messages are at once lost and archived. Today, given the digital revolution, it is possible that television and its sequelae have formed into an invisible matrix of electronic webs that may have heated up beyond all known measures.

Whatever may be the enduring truth of McLuhan's distinctions, the basic premise remains sound—that media are anything but neutral. No one of his time was more like Borges, none more original. Not even Marcuse's very hot *One Dimensional Man,* which was released the same year as *Understanding Media*, stands up against so cool a book as McLuhan's. Still Marcuse and McLuhan spoke to the same telling aspect of media cultures—that they can be powerful and dangerous as they entertain and distract.

The importance of Sam Han's *Navigating Technomedia* is that it provides readers today with an excellent guide and interpretation of the continuing, perhaps more acute, question of how media enhance or debilitate human understanding. Few today would argue that times have changed since, even, the now ancient countercultural days of 1964. People my age are gratefully reliant on young authors like Sam Han and those for whom he speaks who have grown up with—and, we should add, *in*—the new media we late boomers can barely operate.

In my youth, when McLuhan wrote, television was still quite young. It had been but a decade since Elvis shook his pelvis just below the screen Ed Sullivan would give him. But we somehow understood how cool Elvis was. Many of us were children of the 1950s and even if we were less than culturally well informed we somehow knew that cool was in. Then, when the televisual was still in its infancy, jazz entered the generic white consciousness. We were all aware, not only of Elvis, but of the Beat Generation. Only later did we read Jack Kerouac's *On the Road* or listen to John Coltrane. But when (and if) we did, we came to understand that the improvisational road was where we longed to be, which is why the early civil rights and student movements had so much appeal. Cool holds up today as hot does not. We the boomers grew up filling in the details of a world that gave us too little information. But today cool is quite more certainly than those zeros and ones—the on/off nowhere of our mental ennumbering.

What Sam Han accomplishes in this brilliant book is to give readers a clear and succinct summary of the history of media from print through technomedia. What, one might ask, are technomedia? They are nothing less than the super fast, irreal post-televisual media that zip and zap through some ethereal space. This kind of super- or post-cool information can only be understood, if at all, paradoxically. They passeth understanding in that they are not meant to be understood. Or, better, we might say that to understand technomedia like these—media, that is, borne about in the palm, soon to be implanted, one supposes, in the brain—is precisely not to get the point or points they convey; if, that is, these things are true conveyances for signifying messages.

Technomedia mediate not so much by the conveyance of information as by the creation of virtual spaces that are so completely cool as to be off the thermometer of the human mind. We must exercise, if we are to live in or with them, a new kind of critical imagination. Yet, these devices are so evidently both hot and cool, thus neither, purely and simply. The cool freedom they give us lifts us from the terrible heat of life on the ground. In Beijing today, the young live by text messages as they stroll the boulevards, just off which, in the dark, miserable rural migrants toil to build the skyscrapers they will never inhabit.

Sam Han writes of these subjects, from the history of media, to the definitions of technomedia and its own vocabulary, to the new cultural politics of hip-hop and much more. I have, personally, learned more from this little book on this new subject than from any other book I have read. Others—whether new to the field of media studies or old timers trying to deal with them, or the young who are old hands at the keyboard—will learn in *Navigating Technomedia* of the wonders of the new media that may be neither hot nor cool, but decidedly are, somehow, both thrilling and terrifying.

Introduction

> "Modern technology touches on almost everything vital to man's existence—material, mental and spiritual. Indeed, what of man is not involved?"
>
> —Hans Jonas[1]

On Christmas Eve 2005, I woke up in Seoul, my birthplace. Just a few days before, I had flown in from New York (with layovers in San Francisco and Tokyo) on one of my not-so-frequent "return" trips to visit family. Needless to say, after close to twenty-four hours of travel time, I had been tired and in desperate need of rest. So, as is the custom for "guests" in Korea, I ate (and slept) a lot for the first couple of days. But on this day, I needed to wake up and break the cycle of eating and sleeping, not only because my digestive system was beginning to wage struggle, but because I had an appointment to keep.

Having successfully navigated my way through the city's subway system, I got off at the station closest to Sogang University, a prestigious Catholic university located in Seoul, and hopped in a cab. When I pulled up in front of the school's gates, I went straight to the security booth to ask where I could find the School of Social Sciences. Following the directions I was given, I entered the building and went up to the second floor where the sociology offices were housed. I had arrived successfully at the office of Professor Kyung-man Kim.

A month before my trip to Korea, I had made plans with Professor Kim through e-mail. In point of fact, it wasn't even a direct communication. One day, I received a forwarded message from Charles Lemert (the editor of the series of which this book is part), within which he asked Professor Kim, a social theorist and sociologist of science, if he could make some time to meet with me during the time that I would be in Korea. Professor Kim had agreed in an e-mail reply; Charles forwarded me that message as well. By the time I arrived at his office, Professor Kim and I had had one e-mail exchange to establish a date and place: December 24th, his office. The fact that I did not once confirm this appointment with Professor Kim explained the first words (which were in English, I might add) he let out when he saw me waiting for him in front of his office on Christmas Eve: "Are you Sam? I didn't know if you would actually be here."

Upon first glance, Professor Kim's surprise may seem to be rooted in the fact that, for most of us living in an uncertain globalized world, when something goes according to plan, it is an exception. The completion of a stated goal brings with it not simply satisfaction, but also astonishment. Another interpretation of his surprise at my being there could have to do with the very medium by which we made our plans a month prior to the meeting—e-mail. For Professor Kim, making plans through e-mail was not quite as certain as making plans on the phone or in person, in spite of the fact that he teaches and lives in one of the most technologically advanced and savvy nations in the world. As of 2006, 67 percent of the population in South Korea uses the Internet.[2] I mention this to ask if it is not a miracle that all of us in our daily lives, filled with electronic mediations of various kinds—PDAs, iPods, digital cable, webmail, and so forth—are not more often surprised the way Professor Kim was. How is it that we have so easily adapted to new media technologies and managed to incorporate them so smoothly into everyday life?

Those of us living in parts of the world with high penetration of Internet usage are not easily amazed when we conduct the various aspects that make up our everyday lives through e-mail. In institutions of higher education, for instance, we send and receive e-mails from colleagues about meetings of one kind or another that are to take place on certain dates, at certain places and times. More often than not, we negotiate meeting places and times through the dreaded, but pretty well-functioning, "mass e-mail." When these things are agreed upon, those of us whose bodily presence is required at the meetings are not astonished that plans made in the virtual realm are fulfilled in the physical. Though largely unacknowledged in the course of everyday routines, this collapse of the separation between the physical (or actual) and the virtual lies at the core of the present social reality. It manifests itself in the easy familiarity with, or even expectation of, interactions in between the realm of the virtual and the actual, In spite of the fact that many of us are able to remember times before e-mail (and even before the widespread phenomenon of personal computing).

Quite simply, new media technologies allow us to mediate our interactions with the social world and control how we represent our own identities to one another and to ourselves in ways significantly different from the technologies that came before them. More than ever, it is getting to be quite obvious that *technomedia* (a term I will explain later) has brought with it massive social transformations. *Navigating Technomedia* aims to take stock of such changes along several conceptual axes: knowledge, space/time, subjectivity, and politics.

Most people are familiar with the rallying call of the Enlightenment: "Dare to know." It is clear these words of Immanuel Kant have had lasting effect. Notions of the progress of human history and the modern rational individual that came from the Enlightenment continue to linger even amid discourses, roughly described as postmodern, that were said to have deconstructed the "subject." The act of knowing (and accumulating knowledge) remains at the core of our culture.

Two billion people use mobile phones.[3]
More than one billion people use the Internet.[4]
As of May 2007, people worldwide spent an average of 31 hours and 35 minutes per month on the computer.[5]
The total growth in worldwide Internet usage from 2000 to 2007 was 225 percent.[6]

Recently, scholars have described a shift in the economy that began roughly twenty years ago, from a mode of production of manufactured commodities to a mode of information of knowledge commodities.[7] Along with this came corollary modifiers to portray the new society—neoliberal, information, and knowledge. Today, these scholars are likely to find that the realities of the present economy, with its emphasis on the service industry, team-work labor models, and speculative financial investing, corroborate their observations.

A similar but far more alarmed position came from humanist scholars like Neil Postman, who were concerned with information *overload* in the age of television. They must now be quite horrified to see how the Internet has one-upped television in its overloading of information beyond levels of rational comprehension. In either case, none could deny that media technologies were changing our world drastically.

With that said, in the early years of the new millennium, is it a possibility for us to *not* know? To know cannot be said to be a dare any longer. We live in an era in which features such as RSS, XML, and ATOM news feeds, which aim to go beyond the innovation of television's "live-feed" by providing "instant access" to information, are commonplace. The knowledge we gain instantaneously today is quite different from the kind Kant dared us to gain. Media technologies allow humans to become nodes in the network of information flow so that we are not so much accessing data with news feeds, but receiving it much like an electronic interface in a circuit. Technologies are woven increasingly tighter into the social fabric, so much so that, as this book contends, the dichotomy of human and nonhuman, among many others, must be rethought.

The Internet makes clear that humans are not simply the users of technologies. We do not use technologies merely to gain, produce, and accumulate knowledge; technology in the current digital/informational milieu is characterized by *new* knowledge formations. Furthermore, technologies are no longer strictly limited to the sphere of the mind, which the popular and no doubt Enlightenment-oriented concept of "knowledge" implies. For instance, media technologies introduce new and interesting relations with bodies.[8] Though futurists like Howard Rheingold and Raymond Kurzweil believed Virtual Reality (VR) to be the next phase of human-technological mediation, today's bio- and nano-technologies, video game systems, and military technologies demonstrate the already very close link between "reality" and "simulacrum."

As I've said, the current social and cultural milieu can be summed up by one word: information. To escape (or worse, to be excluded from) the mediations around us is inevitably to endure something like a social death. Those of

Martin Heidegger (1886–1976) was a German philosopher most well-known for *Being and Time* (1927), which outlines his philosophy of Da-sein, which in German is idiomatically used synonymously with "existence." For Heidegger, however, "Being" was not enclosed and embodied by a substantive and eternal human "soul." Da-sein instead redefines the idea of existence as determined by a temporal engagement of the world, not a universal one.

us in the global North are not only surrounded by and dependent upon media technologies, but also, as I suggest throughout this book, *of* them. It is apparent today, more than it has ever been, that economic, political, social, and cultural life revolves around technology. The role that technology plays in the everyday is not a new discovery by any means. The use of technology has been the focus of attention for scholars in the history of many disciplinary traditions. Marx was no stranger to the effects of technology on what he famously called the "conditions of production." Freud remarked on the rather unusual nature of telephonic communication with regard to the technique or method of psychoanalysis. More obviously, it is Heidegger who questioned the implications of technology and suggested its connection with human existence or Being in his essay "The Question Concerning Technology" (1949).

Nonetheless, to say that one cannot escape anything technological would be to fall into a determinism that has plagued many critics and theorists of technology during the information technology revolution of the 1990s. Luddites as well as "techies" could not avoid the notion that technology, once embraced and thus made ubiquitous, would lock us into a hyperbolized version of the Weberian iron cage, one which would wholly determine human behavior, for better or for worse. Some critics associated the information technology revolution with globalization, American-style democracy, and neoliberal capitalism, all of which espouse a telos, or fixed end-point, triumphantly marking the reign of market-oriented governmentalities and ideologies. Others, seeing the power of biotechnologies, in particular the genome project as well as the possibilities of stem-cell research, expressed a similarly deterministic view of the perils of technology, as it was no longer merely utilized by human life but able to *create* it.[9] Thus, the information technology (IT) revolution brought with it awareness of matters technological unlike times before, even to the point, as I have mentioned briefly, of extreme polarization of opinions and thoughts on the issue.

Today, in the wake of the excitement of the information technologies revolution, it is difficult to separate what is technology from what is not. The subtitle *Caught in the Web* comes from the philosopher Martin Heidegger's notion of "thrownness." Heidegger argues that human existence—which itself is explored throughout this book—is conditioned by forces beyond the "essence" of the human individual. The individual's existence is caught in the social, cultural, and political conditions of a given time or space. Thus, existence is neither a transcendental nor universal experience, but in fact a spatio-temporally particular experience of being in the world. If we take it seriously, "thrownness" opens us up to critically rethinking the questions that are brought up in the information age, or as Manuel Castells calls it, the "network society." Instead of asking "Is it good or evil?" with regard to technological innovation like many did with regard to television, steering clear of rash value judgments allows us to ask more pressing questions:

1. How have the new media technologies that distinguish the culture of information contributed to the various changes in the world?
2. What are the conditions upon which "information," to use this yet undefined term, can be exchanged, communicated, relayed, dispersed, or, increasingly, copy-and-pasted?

I begin with a history of *technomedia*. It is a term I employ to suggest that "media" and "technology" can no longer be thought of separately from one another. After the influence of the British school of cultural studies in the late 1980s, media studies, along with more traditional academic disciplines, focused on audience receptivity, cultural diffusion, and other social effects of media, but what was missing generally was the study of the media themselves, as well as the technologies that facilitated them. Even Marshall McLuhan, who pioneered the study of media, could not avoid a degree of anthropocentrism—placing the figure of Man as the subject/object of knowledge. Thus, McLuhan's offering of a view of media as merely a prosthesis of humankind is unsurprising.[10] By looking at both the mediating components of information technologies and the dynamics of the technologies themselves, we open up the possibility of exploring, not measuring, social effects of media technologies and also reconsidering the very nature of "the social" or society. As I will argue throughout this book, media and technology, conceptually as well as technically, have collapsed into one another, radically challenging the fundamental concepts of modernity, namely knowledge, space/time, subjectivity, and politics.

In presenting a history of technomedia, I do not follow the leads of more well-known methods of history such as Friedrich Nietzsche's genealogy and Michel Foucault's archaeology.[11] There is no tracing of the history of ideas coupled with epistemological shifts, as in the case of Foucault, nor with a culture of moral valuation, as with Nietzsche. The history of technomedia involves a periodization of media technologies that defies the logic of dominant historiography: linear chronicling of events. Though it is not robust like the scholarship in the *Annales* tradition of Fernand

Braudel or more recently the nonlinear historical work of Manuel De Landa,[12] chapter 1 attempts to discard event-based history by offering periods in which important media technologies emerge linked by similar socio-technological ramifications.

The history of technomedia is *not* a statement of definitive historical events, nor is it a foray into historical, political economy, attempting to study the changes in the technical solutions employed in the labor process in the new economy, which some scholars who study media and technology have already done.[13] Instead, I use these periods—Mechanical, Electric, Broadcast, and Virtual—as strategic placeholders or bookmarks to set up the arguments that come later in the book. By situating these precursors to the Virtual era, of which I believe we are on the cusp, I am attempting to summarize social change that directly correlates to the development of media and technology.

In proceeding from the history of technomedia, we move to the four core themes that situate the concepts throughout the book. I begin with a discussion of knowledge with respect to technomedia, paying special attention to the discourse of the Internet as a medium of "transmission" and "connectivity." Early in its history, much attention was paid to the Internet's infinite communicative potentialities, along with its description as an "information superhighway." These characterizations bring out underlying assumptions within the discourse. Though much of it was generated by those who were selling Internet services, the Internet Service Providers (ISPs), it became a popular scholarly view as well. Many social critics and academics, from all sections of the continuum of political allegiances, began to take stock of this potential. The divide, thus, emerged between those who were excited about new technologies and those who were not. Again, the fault line of this divide was not political allegiance *per se*. In some instances, radical leftists and right-wingers alike were skeptical about these new technologies, while liberal-centrists were shouting cheers for the Internet and a better world that would be more informed and more interconnected as a result.[14] Though both camps seemed to understand that the Internet would be a different type of medium than ones prior—especially television—not much attention was paid to the change in information itself or, in other words, what would *count* as information on the Internet. They failed to see that the fundamental descriptions of communication media were eroding.

Knowledge has undergone a shift on the Internet. The title of chapter 2, "A Rapport with Knowledge," borrows from the subtitle of Jean-Francois Lyotard's eminent *The Postmodern Condition*, which is "A Report on Knowledge." I use "rapport" in place of "report" to draw attention to changes in the relationality of knowledge not only with Truth but also with the notion of *significance*. Lyotard suggests that the postmodern condition is characterized by a suspicion toward the grand narratives of modernity. I extend Lyotard's suspicion onto the view that the Internet is merely a more efficient means of human understanding and knowledge exchange. Technomedia are not neutral; they destabilize the relations of knowledge with not only Truth but also transmission and cultural value. To demonstrate this, I look at blogs as well as websites that play host to bizarre material, which is increasingly characteristic of the Internet. "Viral video" has come to the attention of even television shows that are now playing clips from YouTube and other sources with greater frequency. Knowledge, once transmitted, represented, and expressed utilizing media such as print, has been "informatized," as Michael Hardt and Antonio Negri have put forth in their recent works *Empire* and *Multitude*. It comes now in different and altered forms—in this instance, with regard to viral video and web clips, in unsettled forms known quite literally as bits (or bytes) in the vocabulary of new media. As Bernard Stiegler puts it, knowledge is "discretized"—deconstructed into discrete elements.

Transformations in the way knowledge is transmitted, dispersed, and formed in technomedia must speak to another, concurrent change, namely the relations of space and time, sparking debates surrounding globalization, neoliberalism, and modernity. These debates have come to assume the commonality of the ease of communication across the traditional rules of physics, which of course are grounded firmly in the Newtonian laws of physics. Authors in this vein have commented greatly on the transgressions of rules of physical space and the trumping of time vis-à-vis telecommunications. While acknowledging the importance of these contributions to the rethinking of the role of telecommunications technologies in social life, in chapter 3, I attempt to question the "givenness" or naturalness of space and time.

To investigate the spatial relations of technomedia, I delve into concepts of presence and absence in telecommunications, suggesting that the conditions or context of communication have changed with the Internet. As the foundational conditions for communication have changed, communication itself is altered. Communicating on the Internet makes these changes in the relations of space and time quite clear. On the Internet, time as a measurable constant gives way to *microtempos*—variable temporalities that operate simultaneously. Though they display clock-time, computers run operations in security patches, and software updates for various programs are released by either Apple or Microsoft. Though they are crucial to the upkeep of the computer (much like the oil changes for automobiles needed every 3,000 miles) they are different because there is no space- or time-interval that calls for the update. As I just mentioned, many updates are security updates that are not intervallic but problem-specific. When the ILOVEYOU computer virus wreaked havoc all across the world in 2000, Microsoft scurried to create a security patch to fix the problem.

Media technologies have changed ideals of space. The website, for instance, though presenting itself to be an enclosed space, has embedded in it a series of hyperlinks to other Internet websites. Though it is regulated and enclosed by windows and frames, it is always a meta-space—a space that connects to other spaces. It is a point of contact, a link in a nonlinear chain that engulfs users into architectures of what Gilles Deleuze described as *folds*. In thinking of the Internet as a space of spaces, in which a production of space, as Henri Lefebvre once put it, occurs, we are left with the question: If there is a constant movement of folding and unfolding on the Internet, can there be a measurement of such processes? Moreover, what takes up that space? Scientifically, at least in the realm of elementary textbook physics, matter is associated with volume. It is that which takes up space. We see this most clearly when we think of what are called the *states* or *phases of matter*—liquids, solids, and gases. But for the Internet, what is volume? In the context of technomedia in general, which includes mobile phones,

Blackberries, PDAs, and other Internet-capable devices, what can be said of any object? Does a website take up "space"? Is it simply immaterial?

Precisely because of the technological challenge to the physical limits of the human body and consciousness, critical theories claiming the death of the subject emerged. According to them, the subject could no longer find its source of being in the metaphysical consciousness as purported in the philosophies of Descartes and Kant. Rather, it is characterized by fragmentation and dispersion.[15] Identity, in this vein, is theorized to be inherently unstable. The Internet brings forth a new form of subjective experience precisely because of the changed dynamics of space and time as well as the break from traditional notions of materiality/immateriality. Drawing from both Freud's and Lacan's general theories of neurosis, in chapter 4, I argue that subjectivity on the Internet is defined by social neurosis. Online shopping sites, most specifically eBay and Amazon, two of the biggest gateways for Internet consumption, serve as examples. Neurosis is significant because of its undermining effect on the notions of stable Selfhood that postmodern criticisms have already suggested. Though the analysands (psychoanalytic patients) that were sent to Freud in Vienna were generally labeled "neurotic" or "hysteric" because of perceived abnormalities in behavior, I believe neurosis to a be techno-social condition of subjectivity in a network society. Furthermore, due to this condition of neurosis, and contrary to many scholars who have advanced a view of the Internet as merely a technology of what Max Weber called instrumental rationality, the Internet does *not* entrap humans to a network of relentless productivity or consumption. This two-tailed binary still assumes stability or presence of a subject's physical body and/or consciousness, which I believe to be quite different from how the Internet, in effect, operates. On the other hand it requires us to turn into *interpassive subjectiles*, a dynamic alternative to subject or object that undermines traditional notions of the agency, or instrumental power more specifically, of human "users" of technology. In fact, humans have never been simply the users of technology. This deterministic view of the dynamics of human/technology interaction is couched in the anthropocentric notion that humans are the ultimate creators of technology and masters of nature, thus privileging human presence or participation (either bodily or of consciousness) as a necessary pre-condition for technology's ability to function. Thus, I point to the phenomenon of *lurking* on the Web, when individuals do not make their presence or participation known to others in a given social space such as a blog or discussion board.

With the Internet and various other technomedia challenging traditional forms of subjectivity, the question of how to forge politics begins to come into view. Though this question deserves books unto itself, I group culture, information, and politics together in chapter 5 to suggest the co-constitutive nature of culture, information, and politics of the Internet and technomedia in general. I start with a brief introduction to the thought of Fredric Jameson, who has quite convincingly linked culture and political economy. Jameson's famous argument is that postmodernism is not a subversive knowledge, art, or theory, but, on the contrary, the cultural logic of late capitalism. Those who follow him in their generally post-Marxist ideas, namely Hardt and Negri as well as Maurizio Lazzarato, have identified culture as subject to the forces of production that Marx had once theorized. Drawing from these theorists, I am suggesting an *informational* politics alongside a cultural politics. The panic over copyright infringement that resulted from downloading and file-sharing in general is but one indication of the political nature of a seemingly banal activity of young people on their computers. The properties of a commodity on the Web, such as an mp3 or video file, have in themselves an informational dynamic that allows for its relatively easy reproduction and dissemination. It forces us to reevaluate the contemporary Marxist interpretations of the informational economy by looking specifically into the techniques of reproduction and production alongside conceptions of private property and by thinking about a new type of politics for the information age, in which stable categories of modernity are at best ambivalent in their truth-value.

No matter one's private feelings toward the increasingly visible presence of technomedia in our lives, it is undoubtedly true that they will continue to be so in times yet to come. Technology is not so much here to stay. In many ways, the purpose of *Navigating Technomedia* is to argue that it has always been here, influencing the ways in which we have interacted with, known, and experienced the world and one another.

Notes

1. Hans Jonas, "Toward a Philosophy of Technology," in *Readings in the Philosophy of Technology*, ed. David M. Kaplan (Lanham, Md.: Rowman & Littlefield, 2004), 17.
2. Internet World Stats, www.internetworldstats.com/asia/kr.htm. Maintained by Enrique De Argaez, Miniwatts Marketing Group.
3. CellularOnline, www.cellular.co.za/stats/stats-main.htm, April 2006.

4. Internet World Stats, www.internetworldstats.com/stats.htm.

5. "Global Index Chart: Month of June 2007, Panel Type: Home," Nielsen//NetRatings, www.nielsen-netratings.com/resources.jsp?section=pr_netv&nav=1.

6. "Global Index Chart."

7. See Mark Poster, *The Mode of Information: Poststructuralism and Social Context* (Cambridge, UK: Polity Press, 1990).

8. Most scholarly literature regarding bodies and technologies has come in response, knowingly or unknowingly, to N. Katherine Hayles's *How We Became Posthuman: Virtual Bodies in Cybernetics, Literature, and Informatics* (Chicago: University of Chicago Press, 1999). More recent additions have had, more or less, a Deleuzian orientation by pushing the conceptual limits of "the body" through the study of affect. See Brian Massumi, *Parables for the Virtual: Movement, Affect, Sensation* (Durham, N.C.: Duke University Press, 2002). For a study of technological representations of the body, see Michele White, *The Body and the Screen: Theories of Internet Spectatorship* (Cambridge, Mass.: MIT Press, 2006).

9. See Kaushik Sunder Rajan, *Biocapital: The Constitution of Postgenomic Life* (Durham, N.C.: Duke University Press, 2006).

10. To be fair, it should be noted that Marshall McLuhan follows Norbert Wiener in this theoretical move, which places humans as machines, or at least mechanistic, in themselves.

11. For examples, see Friedrich Nietzsche, *On the Genealogy of Morals* (Oxford; New York: Oxford University Press, 1996). It is worth noting that Foucault too utilized a "genealogy" inspired by Nietzsche in his later works.

12. See Braudel's *Civilization and Capitalism, 15th–18th Centuries*, 3 vols. (Berkeley: University of California Press, 1992) and De Landa's *A Thousand Years of Nonlinear History* (Cambridge, UK: Zone Books, 1997).

13. See Stanley Aronowitz and William DiFazio, *The Jobless Future: Sci-Tech and the Dogma of Work* (Minneapolis: University of Minnesota Press, 1994) and more recently, Stanley Aronowitz and

Jonathan Cutler, *Post-Work: The Wages of Cybernation* (New York; London: Routledge, 1998).

14. Compare, for example, the works of Thomas Friedman, Howard Rheingold, and Neil Postman.

15. See James Heartfield's *The "Death of the Subject" Explained* (Charleston: Booksurge Publishing, 2006) for a concise overview of the key theories regarding the end of subjectivity, in particular that of "fragmentation," which so many theorists have evoked.

1

Technomedia

"Technology made modernity possible."

—Philip Brey[1]

The Modern and Technology

It is common knowledge that the modern or "developed" nations in the world show high levels of technological diffusion. Even within nations, such as the United States, it is pretty obvious from the physical landscape where one will be able to receive a wireless Internet signal or even mobile phone service, and where one will not. There is a greater possibility for the phone call back home to go through in Cleveland or Cincinnati than it is in more rural parts of Ohio. This convergence of technology and progress, as represented in urban geography, exists in the political debates surrounding "the digital divide," a term introduced by Larry Irving, technology adviser to former U.S. President

Bill Clinton. It points to the widening global gap between those with and without regular and effective access to the latest technologies. At the first UN-sponsored World Summit on the Information Society in Geneva, Switzerland in 2003, President Abdoulaye Wade of Senegal called for a UN "digital solidarity fund" to help his country and other nations on the periphery of the world system pay for hardware and software.[2] On top of concerns such as AIDS, poverty, and political unrest, postcoloniality, it seems, necessarily involved a technological oppression, taking form in a structured lag. Hence, for those for whom the promise of globalization had yet to be realized, new media technologies became an integral part of not only the discourse of modernization, but of global justice.

Yet, contrary to the alarm expressed by Wade and the UN, some economists, computer manufacturers, governing bodies (both national and transnational), biomedical companies, and Internet Service Providers, among others, had heralded the ubiquitous presence and increasing affordability of technology as a great stepping stone toward progress, democracy, and unity. Changes in the world—the dot-com boom, the end of the Cold War, and the IT revolution—all taking place within a decade of one another, were reason for many to believe in the prospect of a better and now global society.[3] It was indeed the beginning of a "new world order." The Internet, specifically, brought glimpses of hope for global democratization, in the form of a "true" public sphere that would circumvent traditional political channels. By facilitating more efficient communication among the citizens of the world usually laden with the failures of bureaucracy, the usual stoppages characteristic of bureaucratic institutions could be circumvented. Thomas Friedman and others like him saw the Internet as a media technology upon which a global and democratic collective conscience could be built.

Nevertheless, the liberal dream of progress espoused by Friedman and others did nothing but rearticulate the false hopes of modernity under the guise of technological advancement.[4] From the moment of its mass incorporation, the Internet was an index, not a catalyst, of democratization and modernity, a sign of the (neo)liberal West's triumph over Soviet (Eastern) communism. "Look at our mature democracy! Even our technologies foster democratic practices!" was the chant of liberal critics. Against such optimism, I see the power of information technologies and new media as challenges to these notions altogether. The Internet not only acts as a forum to disperse knowledge but transforms it, hauling into question the certainty of what constitutes knowledge as such. Thus, it supplants its rationalizing potential as per Jurgen Habermas.[5]

Technomedia are media technologies that are defined by interoperability between devices, usually using the Internet as the connective medium. Prior to the popularization of the Internet, media and technology were considered to be distinct entities. Specific media were connected to certain technologies. For example, television could only be watched on television sets with cathode-ray tubes and antennas. Today, television shows are streamed and downloaded all over the Internet through websites like YouTube. Thus, technomedia refers to the fluidity between various media and technologies.

Media technologies, or *technomedia*, do not replace one another but work within each other. This is the first and perhaps most important characteristic of technomedia. As Marshall McLuhan reminds us in *Understanding Media*, the content of one medium always involves another. Thus, "new" and "old" are modifiers that simply do not work; technically, there are no "new technologies." To illustrate, let's take the example of writing as technology, provided by German media theorist Friedrich Kittler. Both alphabetic and pictographic writing are classifications of speech.[6] Kittler notes that speech itself is a technology of translating "everyday languages"—or thoughts—into a formal system of phonemes. Writing absorbs speech, which earlier integrated thought. Thus, the evolution of media technologies is nonlinear in character, consisting of incorporations of prior incorporations, foldings and unfoldings of form that Lev Manovich calls "meta-media."

Today, technomedia make even clearer that specific media are not tied down to a corresponding technology. We can take a look at the electronics industry, which relies heavily on the multitasking capabilities and interoperability of their products. Take the most basic of examples, the mobile phone. Clearly, it is not just for phone calls anymore. Instant messaging, SMS (short-message service, more commonly referred to as text messaging), mp3 playback, camera functions, and Internet access are basic for most cellular plans across the world. The Internet (and of course personal computing) is the clearest illustration of technomedia. Having come a long way since its nascent stages as a product of DARPA (Defense Advanced Research Projects Agency), the Internet has become a progressively more inclusive media technology, one in which

Marshall McLuhan (1911–1980) is considered the pioneer of media theory in North America. Known, for the most part, for his declaration that "the medium is the message," McLuhan laid out the foundations of the study of media, what he called "media ecology," in his most popular work, *Understanding Media*, a sweeping overview of the effects of media on human consciousness. His insistence on stripping media technologies of their "neutrality" fueled his formulation of media as differentiated into "hot" and "cold." He was most famous for arguing in *Understanding Media* and *Gutenberg Galaxy* that media have an effect on human perceptions.

Hot/Cold Media: Hot media is already full of detailed information at the disposal of the human user, so to speak. Books, photography, and radio are examples of hot media. Cold media is not as saturated with detailed information, thus requiring humans to engage actively in order to receive whatever is transmitted through it. He gives the examples of television and the telephone as cold media.

technological developments of past and future somehow incorporate one another. Even the newest generation of video game consoles—XBOX 360, Nintendo Wii, and PlayStation 3—boast of their networking capabilities.[7] We can see that connecting to the Internet is not strictly limited to central technological gateways of access. The Internet is a media technology that allows for, and relies on, other technologies to connect to and utilize it from nearly anywhere, posing for us a radically different look at what technology and media are, as well as how they operate.

A History

When McLuhan famously wrote that media were the "extensions of man," the externalization of the human central nervous system, he did not imagine that technology could exert its power over us in the way that the Internet and other technomedia have. The Internet's power and reach calls for a reperiodization of the history of media and technology, one in which the linear chronicle of "which technology was invented when" is abandoned for a look into the threads of similarities and differences among technologies in terms of their social effects, as well as a consideration of the disintegrating separation between technology and media. There are three eras—Mechanical, Electric, and Broadcast—leading up to the Virtual, of which I believe we are on the verge.

Mechanical

According to many scholars, print emerged in Korea sometime between 1234 and 1241. Gutenberg developed the movable-type, alphabetic metal press in 1440, setting the stage for mass dispersal of information across Europe.[8] The invention and mass incorporation of print features an important social effect of media. Print is an extension and renovation of writing. It catalyzed the obsession with standardization that would later appear in the culture of modernity. With print's dependence on a causal logic, the sequential phonetic alphabet, exact reproducibility was possible. As McLuhan astutely noted, printing set the groundwork for homogeneity, uniformity, and continuity,[9] which it inherited from Newtonian physics and Euclidean geometry (more on this to come).

McLuhan's once-student and oft-interlocutor Walter Ong, in his monumental study *Orality and Literacy*, argues that alphabetic (as opposed to pictographic) writing systems provided the foundation for what McLuhan once called, rather ungracefully, "typographic man."[10] Ong locates the key to the typographic tendency in human consciousness and perception in the alphabetic system's ability to break up "the word into spatial equivalents of phonemic units."[11] Alphabetic letterpress printing technology made letters exist (physically) prior to their utterance or articulation. As Ong notes, this characteristic of the printing press, with each letter having its own metal type, "embedded the word itself deeply in the manufacturing process and made it into a kind of commodity."[12] Before communications media, there was a separation between internally conceived and perceived ideas and their exteriorized articulation in the form of a linguistic medium, usually speech or script. Print technology marked a drastic shift in interhuman communication and perception through its mechanized rationalization of *parole* to *la langue*, disturbing the ontological privilege traditionally given to ideation and mental processes.[13]

Another effect of print that Ong highlights is what he deems "sight-dominance."[14] Though the emergence of writing did indeed plant the seeds for the modern obsession with the visual, print solidified the end of the "oral-aural" culture of premodernity by "lock[ing] words into position in this space."[15] Mechanized rationalization or commodification were now coupled with the development of a sensory hierarchy in knowledge distribution that shifted away from orality/aurality towards visuality. Print allowed for what was visible, detected by human eyes, to be the standard for legislating "correctness" or accuracy. In oral/aural cultures, ideas were always, in some ways, under negotiation. Philosopher K. Anthony Appiah has written about the differences between the West African sensibility of language and speech (and, I would add, knowledge and culture) as shared, in contrast to European ideals of individualized ownership.[16] In much the same way, Ong argues that print

> situate[s] utterance and thought on a surface disengaged from everything else, but it also goes farther [than writing] in suggesting self-containment. Print encloses thought in thousands of copies of a work of exactly the same visual and physical consistency.[17]

In these ways, print combined the logic of mechanical reproductivity in its pioneering of standardization techniques with the triumph of the visual over the oral/aural.

Louis J. M. Daguerre completed the development of daguerreotype photography in 1839. Though the daguerreotype differed from modern photography in that it did not require a negative (the chemical process bears similarities to Polaroid pictures), its function was quite the same. The daguerreotype camera captured and *stored* isolated moments in space and time, producing a mirroring effect of the external world. With the invention of celluloid film or film-stock, the camera emerged as the recorder (and *recoder*) of experience by the late nineteenth century, not quite superseding the written word, but challenging its place as the sole bearer and storage of human experience, extending the visual turn that characterized the onset of print on a mass scale.

Cultural critic Susan Sontag, in an interview for *The Boston Review* conducted after her famous essays on photography were published, identifies two brands of photographers—moralist and scientist.[18] Moralist photographers cover "war, poverty, natural catastrophes, accidents," what Sontag describes as "disaster and decay." Photographers of the scientific persuasion, on the other hand, she suggests, are discoverers of beauty in the "random and banal." The world, to the scientific photographer, becomes an archaeological site to do the hard work of digging beneath the surface. "It's a way," Sontag says, "of aestheticizing the whole world."[19] The culture of photographic aesthetics is crucial to the procurement of the "photographic sensibility" of the post–World War II consumer society. Us late- or postmoderns have a sense of the "picturesque," which is a "method of appropriating and transforming reality—in pieces."[20] Thus, what is seen—the visual—which rose to importance with the onset of print, is transformed to the photographic.

Sontag's criticism is of particular value in the social and cultural history of technomedia due in part to the fact that she addresses similar themes as Walter Ong in his assessment of the effects of print technologies. What Sontag theorizes as the photographic sensibility in Western, consumer-oriented visuality contains a layer of commodification on top of the one identified by Ong. "Photographs," she says, "convert works of art into items of information. They do this by making parts and wholes equivalent."[21] This other commodification that I refer to is quite simply the intrusion of the photographic frame into the human understanding of ocularity. Therefore, the contents of the photograph, what in technical terms is called "composition," no longer are stable objects unto themselves as Kant would have liked. They are stripped down and mechanized into informational bits.

Early film technology extended photography's use of celluloid film, and by the 1880s, the motion-picture camera was in use. In 1896, the Lumiere Brothers began to show motion pictures around the world. Their first film *La sortie des usines Lumière* (1895) captured the relatively mundane (by today's standards) activity of workers leaving a factory. This detail is not to be overlooked. The history of film and the film industry, more than any other media, runs parallel to that of industrialization, modernity, and capitalism. The images captured by the Lumiere Brothers "documented" (it is commonly believed that the exit of the workers from the factory was staged—though I imagine the workers were playing themselves) the rise of industry, in particular the rise of the factory as the center of economic and social life.

This fact—or, better yet, staged fact—brings us to a point where we see the emergence of early film's claim to truth in the West's culture of knowledge. The projection of moving images provided mediation of an empirical, and of course visual, representation, what today would most likely be called documentary or cinema verite. The function of the video camera is simply to capture what is real, like its still cousin. Yet, film was distinct from photography because it was able to capture experience *in sequence,* which photography could not do. It recorded motion and emulated motion through camerawork—cinematography. For instance, the pan- and dolly-shot recreate bodily motion as experienced by human eyes. In other words, camera work in film moved beyond the still image context of photography to mirror human sensory experience *tout court* in the medium.[22]

Electric

Telegraphy became widespread in the 1830s, starting in the United Kingdom. Telegraphs proposed the possibility of written communication no longer bound tightly by physical space. Though telegraphs required cables called transmission lines, they signaled a shift from letter writing. Letters required physical transport every time a communication was to occur, which entailed high amounts of labor power as well as transport costs. The telegraph, on the contrary, allowed the exchange of messages with a sharp decrease in labor power and an increase in efficiency. In addition, the telegraph marked a symbolic moment in which information dispersal methods began to change gears from mechanical reproduction to electricity. The printing press, embedded in its mechanical logic, could not match the speed at which information could move with telegraphy. Though books and other printed materials were able to pack more "information," as in more words in a given space, the telegraph's novelty came from its ability to bend time and space. Though what McLuhan called the electric age would come much later, the telegraph pushed the speed of everyday life.

In 1875, Alexander Graham Bell famously made a transmission allowing two users to converse in *real-time*. The telephone amplified human faculties allowing the unity of two functions: ear and voice. This would not happen in film—in the form of sound-pictures—until many years later. The telephone mirrored human conversation, acting as a discourse machine. It was the first bi-directional communications technology of its kind.

Avital Ronell uses the figure of *telephony* to rethink the notion of communication beyond the model of exchange or sharing of information in *The Telephone Book*. No stranger to the thought of Jacques Derrida and Hélène Cixous, Ronell makes a claim for what Franc Schuerewegen calls a "telephonic logic,"[23] or a telephonocentrism, in all types of communication in contemporary society.[24] She presents telephonic communication as representative of a ghostly voice, what Derrida labels "hauntology."[25] The telephone, in essence, facilitates the ghostly effect of communication disconnected from corporeal contact. Ronell argues that all forms of communication are based on an incessant *telephonage*, an alienating effect that undermines the traditional interpretation of dialogic or equitable communication based on the presence of communicable entities, what in Enlightenment terms would be called *subjects*.[26] Therefore, she describes telephonic communication, or electric speech, as schizophrenia, which "gives us exemplary access to the fundamental shifts in affectivity and corporeal organization produced and commanded by technology."[27]

By the early twentieth century, innovations in editing and storytelling techniques converted films to cinema. The mass medium of cinema developed its own modes of production and distribution as well as its own system of aesthetics and narrative. Dutch media theorist Sven Lutticken argues that cinema inherited strategies from the nineteenth-century culture industry, most especially the narrative form of serials and mass-circulation novels that developed plot in a linear, chronological fashion.[28] As cinema grew less hardwired to its capturing technology—the camera—it was able to narrate in ways other media preceding it could not. Undoubtedly, many filmmakers stuck to the narrative formulas of traditional film grammar, as laid out by D. W. Griffith. (His film *The Birth of a Nation* (1915) is credited with laying out the foundations of cinematic narrative.) Yet, cinema was also the medium utilized by those at the forefront of experimenting with the limits of narrative. Alfred Hitchcock, Sergei Eisenstein, and Luis Buñel, just to mention a few, created films that stretched the limits of subjectivity and narrative as descended from the novel.[29]

Hollywood was the center of what Theodor Adorno and Max Horkheimer famously dubbed "mass culture,"[30] the convergence and flattening of "high" and "low" culture. In the Hollywood system, a handful of studios, quite literally called "the Big Five," controlled the production and distribution of films in the United States. These studios were the precursor to the types of media conglomerates seen today. The Studio System, as it would be called in film history, pioneered the "broadcast" model, characterized by a centralized locale of production and a dispersed system of distribution and reception.

The personal computer had its start (as did ARPAnet, the predecessor of the Internet) at the Department of Defense's Advanced Research Projects Agency (ARPA), which was formed in direct response to the Soviet launch of Sputnik in 1957. ARPA scientist J.C.R. Licklider was one of the first to champion the possibilities of human-computer interactivity.[31] Though his ability to see the future of what was then rather crudely called "man-machine symbiosis" made him a pioneer in the agency for which he worked, Licklider was not alone in supporting the importance of studying technologies to forming a theory of society.

Norbert Wiener, who coined the term *cybernetics*, in a similar vein notes that a study of society necessarily includes a "study of the messages and the communication facilities which belong to it."[32]

One of Wiener's many major contributions to thinking about technology and society is his insistence on the mechanical nature of all organisms. In essence, he considered human physiology as technological. There is no better illustration of this than his concept of *feedback*. Wiener argues that all organisms have a method of learning, which he theorizes as an enclosed communications system. Feedback, he suggests

> is a method of controlling a system by reinserting into it the results of its past performance. If . . . the information which proceeds backward from the performance is able to change the general method and pattern of performance, we have a process which may be called learning.[33]

Though he did not live long enough to see a machine fully capable of these ideas, Wiener did conceive of something close to what we have today in the form of the computerized technomedia, which is capable of both transmission/reception and production of information.

Broadcast

Before radio broadcasts as we have today (which are themselves undergoing a change with the competition from podcasts, Internet, and satellite radio) came about, there was of course radio technology similar to walkie talkies, which was used predominantly by ships at sea to communicate with one another. With the advent of broadcasting, radio became far more centralized, becoming one of the first media to experience the centralizing force of the broadcast model. Generally, radio, as it was popularized in Europe and the United States, meant listening, not transmitting. Information flowed only one way. It was the quickest method of spreading information and quickly became the dominant cultural fixture in homes. Everything from serial radio soaps to presidential addresses was broadcast over radio waves.

Paul Nipkow, from Germany, created the first electromechanical television system in 1885. However, the prevalent version of television (with cathode-ray tubes) came into effect at the beginning of the twentieth century. Broadcast television came about afterward, in the late 1920s. After World War II, along with the baby boom came a TV boom. TV sets were a mainstay, replacing radio, in American culture. Indubitably, the living room became the TV room. The nightly news and prime-time programs were the method by which cultural dispersion could occur at the quickest pace. Arguably, Beatlemania and even the Civil Rights movement in America could not have occurred without the medium of television.[34] With the introduction of cable television, the once-limited option of viewers to flick back and forth between the major networks became a thing of the past, with the most basic cable television packages offering 70 channels or so.

Richard Dienst, in his critical assessment of Raymond Williams's ground-breaking analysis of television, argues that it expanded the world. "The dreams of television," he writes, "were already global . . . televisuality was immediately imagined as an all-encompassing putting-into-view of the world."[35] He traces an important parallel between the televisual broadcast system, "a social space [achieved] through the distribution of transmitters and receivers,"[36] and the trajectory of capitalism or even modernity itself.[37] "Television," he argues, "first offered a transparent *world*."[38] He suggests that there is a convergence and conflict of ideologies on television—those of the state, culture, and inevitably the market. It is also a "new combination of what used to appear as separate practices—the levels of market-based economics, technological deployment, and ideological representation."[39] Consequently, television is a hodge-podge of social forces, as well as a mixture of formal techniques old and new. Additionally, Dienst's criticism brings out key formal and technical differences between film and television. Unlike that of cinema, the televisual image is always subject to static. "With transmission," he writes, "images and sets of images pass the time and fill out the current: in this sense television is always 'live.'" In its tentative connectivity, television differs from film, which necessarily disconnects the "here-and-now" viewing of the film by the audience from the "then-and-there" production of the film. Due to such a difference, Dienst theorizes television as not a producer of images like cinema, but rather a platform for what he calls *fields of visuality*.[40]

Changing Relations

In each of these periods, the relations of space and time changed significantly, as did knowledge and subjectivity. The technologies of the mechanical era emphasized the importance of exact, mass reproducibility of not only commodities but also human experience. Print's model for mass distribution provided similar methods of reproduction and distribution for photography and early film. As with the contemporaneously burgeoning factory system, reproduced versions of any original could be made with increased speed and ease. The innovations of the electric age ripped open the seam that once bound the distribution of information to the limits of physical space. With the ability to transmit information through electrical lines, the exchange of information could occur in real time. The broadcast era represents the explosion of media to a level of mass usage. It is also important to mention that media of this era brought forth the invention of the "live feed," an extension of the real-time capabilities of the technological innovations of the electric age. With this unprecedented ability to communicate across vast areas of space, radio and television became fixtures in everyday life and acted as the main sources for news, information, and entertainment.

These changes allow us to scrutinize what it is that technomedia do. Are technomedia means of communication? If so, what do they communicate—human ideas or just "information"? Furthermore, are "information" and human ideas mutually exclusive? These are questions that Friedrich Kittler attempts to answer when he critiques the philosophical understanding of "communication," claiming that it is too general. According to Enlightenment philosophers, communication is "the rendering into speech of perceived ideas and consequently the linking of isolated individuals through . . . language." They fail, Kittler writes, "to enquire how, without language, people are supposed to have arrived at their ideas and conceptions in the first place."[41] To resolve such a dilemma, Kittler reminds us of the distinction upheld by early information theory and systems theory between *information systems* and *communications systems*. Information systems have the potential for data storage, processing, and transmission whereas communication systems (media) are thought to deal with control of signal traffic.[42] However, as he notes, this distinction has collapsed because contemporary technomedia are able to transmit and store data, while simultaneously being able to control commands through algorithms.[43] Taking a cue from Kittler, I assert that technomedia is unlike prior media and technologies precisely because of its ability to break down and reconstitute the classic categories of knowledge, space, time, subjectivity, and politics upheld by philosophy, information, and systems theory. It challenges thought and pushes it toward the future.

Notes

1. Philip Brey, "Theorizing Modernity and Technology," in *Modernity and Technology*, ed. Thomas J. Misa, Philip Brey, and Andrew Feenberg (Cambridge, Mass.: MIT Press, 2003), 33.
2. Alfred Hermida, "UN Summit Pledges Net for All," *BBC News Online*, December 12, 2003, at news.bbc.co.uk/1/hi/technology/3314921.stm.
3. Critics such as Thomas Friedman authored many articles and books in the 1990s pointing to the hope that new communications technologies would give rise to a potential for democracy across the world through neoliberal globalization. See Thomas Friedman, *The Lexus and the Olive Tree*, 1st ed. (New York: Farrar, Straus and Giroux, 1999). On the other side of the same coin, conservative figures the likes of Francis Fukuyama and Samuel Huntington saw the fall of Communism as the triumph of American-style democracy. See Francis Fukuyama, *The End of History and the Last*

Man (New York: Avon Books, 1993) and Samuel Huntington, *The Clash of Civilizations and the Remaking of World Order* (New York: Simon & Schuster, 1996).

4. It must be noted that Friedman has altered his position with regard to the hope for democracy through technology in his more recent work. See *The World Is Flat: A Brief History of the Twenty-First Century*, 1st ed. (New York: Farrar, Straus and Giroux, 2005).

5. A more detailed discussion of Habermas's "communicative rationality" appears in the next chapter.

6. Friedrich Kittler, "The History of Communication Media," *CTHEORY* July 30, 1996, at www.ctheory.net/articles.aspx?id=45.

7. Though these three consoles differ in many ways from one another, they all share the capacity for users to connect to play against others online.

8. Though the Korean printing press did indeed have movable type, it did not have separate letters. At the time, spoken Korean differed from Chinese, but written Korean borrowed heavily from Chinese characters; this system was called *hanja*. The modern-day Korean phonemic alphabet, developed under the rule of King Sejong, was introduced in the 1440s. Thus, the Korean press of the thirteenth century was quite different from Gutenberg's.

9. Marshall McLuhan, *Understanding Media: The Extensions of Man* (New York: McGraw-Hill, 1964), 89.

10. The subtitle of McLuhan's *The Gutenberg Galaxy* (Toronto; London: University of Toronto Press, 1962) is *The Making of Typographic Man*. Though he was not influenced by deconstruction, McLuhan, in this book, attempts to identify a centering force in print technology that could be compared to what Jacques Derrida famously labeled "logocentrism."

11. Walter Ong, *Orality and Literacy: Technologizing the Word* (London: Routledge, 1993), 116.

12. Ong, *Orality and Literacy*, 117.

13. The Swiss linguist Ferdinand de Saussure, considered the founder of modern semiology (the study of signs), is responsible for the benchmark text of structuralism: *The Course in General*

Linguistics, published in 1916 (Chicago: Open Court, 1998). In it, he argues that *parole* is made up of differentiated parts, whereas *la langue* is a system of signs attached to meanings—social, cultural, or otherwise.

14. Ong, *Orality and Literacy*, 121.

15. Ong, *Orality and Literacy*, 121.

16. See his *In My Father's House: Africa in the Philosophy of Culture* (New York: Oxford University Press, 1992).

17. Ong, *Orality and Literacy*, 132.

18. Susan Sontag and Geoffrey Movius, "An Interview with Susan Sontag," *The Boston Review* June 1975.

19. Sontag and Movius, "An Interview."

20. Sontag and Movius, "An Interview." For a commentary on Sontag's theories on photography, see Jay Prosser's opening remarks for *Picturing Atrocity: Photography in Crisis*, a conference/webcast sponsored by the Center for Humanities at the Graduate Center of the City University of New York and the University of Leeds, UK, held on December 9, 2005, at www.photographyandatrocity.leeds.ac.uk/pa_00/pa_00.htm.

21. Sontag and Movius, "An Interview."

22. "Experience," as such, is up for discussion. Many poststructural critics have taken to task phenomenology's notion of it. My line of thought posits that technomedia has as much to do with influencing understandings of human experience as it has with reproducing it. Here, I am using such a loaded phrase in order to argue that the rather homocentric or anthropomorphic personification of technologies existed as intellectual dominants of modernity.

23. Franc Schuerewegen, "A Telephone Conversation: Fragments," *Diacritics* 24, no. 4 (1994): 35.

24. The allusion to Cixous and Derrida is in reference to "phallogocentrism" and "logocentrism" respectively, which Ronell, once their student, has drawn upon.

25. See Jacques Derrida, *Specters of Marx: The State of the Debt, the Work of Mourning, and the New International* (New York: Routledge, 1994).

26. I discuss the notions of absence/presence inspired by Jacques Derrida in chapter 3 on space/time.

27. Avital Ronell, *The Telephone Book: Technology—Schizophrenia—Electric Speech* (Lincoln: University of Nebraska Press, 1989), 109.

28. Sven Lutticken, "Suspense and . . . Surprise," *New Left Review* no. 40 (2006): 97.

29. See Slavoj Zizek, *Everything You Always Wanted to Know About Lacan (But Were Afraid to Ask Hitchcock)* (London; New York: Verso, 1992). Zizek's Lacanian readings of Hitchcock point out the unusual nature of subjectivity in many of Hitchcock's films.

30. Though the "culture industry" is Adorno and Horkheimer's term, I would say that nearly all of the scholars associated with the Frankfurt School did theorize such a convergence. One famous example is Walter Benjamin's "The Work of Art in the Age of Mechanical Reproduction," in *Illuminations*, ed. Hannah Arendt and Harry Zohn (New York: Schocken Books, 1968). For Adorno and Horkheimer's definition of the term, see *Dialectic of Enlightenment* (London: Verso, 1997).

31. For a short, interactive history of multimedia, see www.artmuseum.net/w2vr/contents.html.

32. Norbert Wiener, *The Human Use of Human Beings: Cybernetics and Society* (New York: Avon Books, 1967), 25.

33. Wiener, *Human Use*, 84.

34. Todd Gitlin argues that media coverage contributed to both the rise and fall of the American Left by determining the modes of reception. See *The Whole World Is Watching* (Berkeley: University of California Press, 2003).

35. Richard Dienst, *Still Life in Real Time: Theory after Television* (Durham, N.C.: Duke University Press, 1994), 6.

36. Dienst, *Still Life*, 7.

37. To see the connection between globalization, capitalism, and modernity, see works by David Harvey, specifically *The New Imperialism* (Oxford; New York: Oxford University Press, 2003) and *A Brief History of Neoliberalism* (Oxford: Oxford University Press, 2005). I also suggest the reader to go to the early work of Immanuel Wallerstein, especially the three volumes of *The Modern World System* (New York: Academic Press).

38. Dienst, *Still Life*, 11.

39. Dienst, *Still Life*, 13.

40. Dienst, *Still Life*, 20.

41. Kittler, "A History," 1.

42. Kittler, "A History," 1.

43. Kittler, "A History," 2.

2

A Rapport with Knowledge

"All knowledge which gives power is technology."

—Friedrich A. Kittler[1]

The Unreason of Technomedia

In a very famous anecdote from the pages of *Mythologies*, French literary theorist and critic Roland Barthes, while sitting in a barbershop, sees a copy of *Paris-Match*, a popular French weekly. On the cover is a young man of African descent "in a French uniform . . . saluting, with his eyes uplifted, probably fixed on a fold of the tricolour [the French flag]."[2] This picture, says Barthes, signifies a string of ideological assumptions. It says

> that France is a great Empire, that all her sons, without any colour discrimination, faithfully serve under her flag, and that there is no better answer to the detractors of an

> alleged colonialism than the zeal shown by this Negro in serving his so-called oppressors.[3]

Barthes is arguing that a cultural product such as a photograph on the cover of *Paris-Match* does not simply depict or capture a neutral (value-free) moment or event in space and/or in time. Underneath the photograph is the erasure of the history of colonialism. In fact, Barthes argues that print procures a particular type of reason, but one officiated by a historical and cultural memory that utilizes images and narrative to constitute what Foucault would call an *episteme*, which dictates what is and is not thinkable. Reason, in Barthes' view, is not a critical function of the life-world, so to speak, but a product of the relations and structures of power.

How do we who use various media in our everyday routines think of media? For some, media is a technology that represents the actual in its purest form. News coverage on television, radio, and the Internet, for instance, purports to represent lived moments, grounded in a material reality. These events are condensed and compressed electronically for those of us ("rational" subjects, no doubt) who were not present physically for the moment or event. (In Disney World, there are even places to take pictures called Kodak PhotoSpots, strategically located to produce a representation of a "lived moment" of a family enjoying themselves.) In this representational ideal of technomedia, there lies an assumption that an unhindered, pure transmission of information, from subject to subject and also from media to subject, is possible. Media in this framework resembles Habermas's "ideal-speech situation," by which, under Reason's umbrella, men could come together one day to reach the harmony of the Hegelian state of Absolute Knowledge. The Internet, however, shatters this.

As I mentioned, technomedia always distorts and translates. The Internet makes clear the *impossibility* of a "field of equivocality," or the ideal-speech situation, because of the many distractions that come with the territory. To anyone who uses the Internet on a regular basis, junk e-mail ("spam"), pop-up ads, and unsolicited sexually tinged instant messages are, if not regular, familiar. The pop-up ad in particular exhibits the trickiness of the Internet quite well. Many pop-ups advertise opportunities to win cash or prizes (cruises, iPods, and even computers occasionally). As most Internet users know, these are scams. Even with new pop-up-blocking features on popular web browsers, the occasional pesky pop-up unexpectedly appears. Moreover, every Google search has ads that appear based on the words typed into the search box.

Those who have ever had the experience of accidentally clicking on a pop-up or moving their mouse over the advertisement banners on top or on the sides of a website know it is quite an experience. Most pop-ads have a *sui generis* system of regeneration. One misled click on a pop-up could lead to a flooded computer desktop full of them. Though it seems to have the potential to be a platform for a new rational public sphere, the Internet currently plays host to what seems like chaotic randomness.

That some thinkers still hope that a rational public sphere will come about through unhindered media points to the fact that modern categories of knowledge are in crisis at a time when it is becoming clear that media technology is not simply another instrument of rational truth and representation. We still speak of media as if there is reportage when there cannot be; the prevalence of distortions or, as I like to think of them, *tricks* of all sorts must be recognized. The Internet queers the notion that the world is just there to be observed and understood by humans. While remaining a communication medium, the Internet has distractions built in.

The Technological Conditions of Humanity

According to classical anthropology, technologies are tools for understanding and living off nature. Quite simply, they make human subsistence possible. Modern technologies, in turn, expanded the project of human survival to something very different: the pursuit of complete mastery of nature. The project of survival thus gave way to the humanist project of *exploiting* nature for the advantage of mankind. Nature as object, human as subject—this dichotomous formulation grounded the move toward scientific knowledge that technology took in the modern era. Technology would make visible and knowable that which was not. Within the confines of the linear trajectory of human progress, nothing would be left to the imagination. The closer humans were able to get to Truth, the better off they would be. Technology was the engine on the vehicle of science, which was on the path to the unquestioned truth of nature. Modern technology proved that nature had been conquered and colonized by mankind. This was one of the main criticisms of Marx regarding modern industry: It destroyed the dialectical exchange of man and nature. The factory, steadily jockeying for position as the locus of social life in capitalism, was the gateway from which technological advances cast the widest net. The onset of the new mode of production and its alienated form of wage-slavery first showed signs of its power and viciousness in the workplaces of the proletariat. In the name of increased productivity, technology in the factory made everything faster and easier, changing the relations between nature and the worker's body and truncating, among other things, the limits of space and time that had characterized feudal life. It began not only to reveal nature, but also to violate its laws and dominate it.

What was revealed, as historian Anson Rabinbach argues, was energy. Nineteenth-century physics "discovered" the universal constant of matter as energy, which could be converted into various forms.[4] Following this discovery, the laws of thermodynamics asserted the primacy of energy, often in the form of heat, as the basis for all mechanical work. The factory was the basis for a scientific materialism that rested upon energy as "the quintessential element of all experience, both organic and inorganic," dedifferentiating society and nature to an indistinguishable point.[5] *Kraft*, or energy, was thought of as the transformation of society into a mirror of nature. The image in the mirror was reminiscent of Lacan's *imago*, the image-form that gives meaning to the child in the mirror stage. In this scenario, the mirror of nature was the thermodynamic social body that labored. The distinctions between humans and machines were, from the beginnings of industrial capitalism, already being blurred.

Inescapable Mediations

British sociologist John B. Thompson suggests that mediated communication is a product of a distinctly modern project. "The development of communication media," he writes, "was interwoven in complex ways with a number of other *developmental* processes which, taken together, were constitutive of what we have come to call 'modernity.'"[6] He argues that the transition from feudalism to modernity necessarily included a process of mediatization, explaining that modern communication media replaced traditional face-to-face interactions as the dominant forms of greater social interactions. Thus, modernity had altered the form of social relations. To make this argument, he refers to Jurgen Habermas's renowned theory of communicative action, which links the circulation of printed materials in early modern Europe to the formation of the public sphere in modern societies. The widespread acceptance of print media (newspapers, pamphlets, magazines, etc.), for Habermas, indicates the entrance of media *into* daily life (what he frequently calls the "life-world") and contributes to, as he notes, a public sphere predicated upon collective rationality. The public sphere, as defined by Habermas, guarantees access to all citizens with the freedom to confer in an unrestricted fashion.[7] Print media, he argues, specifically newspapers and (moralistic and critical) journals, initiated dialog and discussion, giving birth to "public authority."[8] These particular media, according to Habermas, legitimized the transformation of authority from monarchical into "rational," procured from the supposed collective rationalism of the public.

Habermas's vision of the enlightened or "rational" bourgeois public sphere is a response to what he calls the "refeudalized" public sphere of today, characterized by the "peculiar weakening of its critical functions."[9] He argues that we must work to recover the rational public sphere by

Jurgen Habermas (1929–) is a leading German social theorist in the tradition of the Frankfurt School of Critical Theory. He studied at the Institute for Social Research in Frankfurt under Max Horkheimer and Theodor Adorno and later went on to direct the institute until 1993, after a stint as director of the Max Planck Institute. He is author of more than twenty books. Some of his most well-known publications are *Theory of Communicative Action* (1981), *The Structural Transformations of the Public Sphere* (1989), and *Knowledge and Human Interests* (1968). In many of his writings, he defends **the public sphere** and civil society as hopeful seeds for the procurement of democracy beyond nation-states. Habermas defines the public sphere as a democratic space for the exchange of ideas and information, identifying the coffee houses and salons of eighteenth-century Europe, where bourgeois individuals would engage in discussions on art, literature, and politics, as an example. This is an ideal type of "**communicative rationality**," a dominant Reason that allows for the catapulting of one's own personal opinions and thoughts onto a public political agenda.

returning to a medium like print, with critical functions. This process, he argues, will restore the public sphere for democracy. This restoration, Habermas says, will encourage rival organizations, though themselves fighting for control of the public sphere, to commit to the rational reorganization of power in social and political structures. It is more than clear that he has yet to lose hope in the State, despite the visible failures of even the social democratic welfare states like France, which has dealt with some of its own incapacities to carry out the "incomplete project of modernity" for *all* of its citizens.

Technomediated Worldviews

In *The Order of Things*, Michel Foucault presents his notion of the *episteme*, arguing (and I am oversimplifying slightly) the presence of a conditionality of acceptable knowledge discursively constituted by an era's conception of Truth.[10] Similarly, the technologies and media of a given era influence the scope of the body of knowledge as well as the expandability of the human universe. As McLuhan noted, movable-type printing was important in the construction of not just a view of our world, but a conception of our galaxy.[11] In *Understanding Media*, he offers a vision of the future in which the electrical connections between humans would bring people together in the name of universal humanity, coining the now famous term "global village." Although he seems to be quite excited about digitalization and its future, McLuhan's "village" implies a series of meanings that are stuck in tradition and nostalgia. For him, the massive mediations of the world entail a return to social relations grounded in "authentic" human relations. In direct contrast with the analyses of the world as entrenched in capitalist social relations offered by Weber and Marx, he envisages a world in which there is "no longer a specialist explosion of increasing alienation."[12] The global village technologically extends "to involve us in the whole of mankind."[13] Whereas Marx's figure of *alienation* and Weber's *iron cage* indicated a rather bleak outlook for the individual in a world wrought by capitalism, McLuhan, though not speaking about capitalism directly, sees an end to this fragmentary way of life with the birth of a unified one in the electric age. In the global village, technomedia are hegemonic, spreading a "faith that concerns the ultimate harmony of all being."[14]

Martin Heidegger was far less hopeful about technology. Like McLuhan, Heidegger cautions against viewing technology as "neutral," warning us of a methodological "blindness"[15] that sees technology as "instrumental or anthropological."[16] He identifies this as the link between the ancient doctrine of technology, inherited from the Greeks, and the modern one of scientificity. This instrumentality of technology, both ancient and modern, assumes a value-free causality; Heidegger notes that this resembles the authority of Adam to use nature at his disposal. In addition, it is rooted in what he deems a "bringing-forth," defined as "bringing unconcealment out of concealment."[17] He delineates a clear distinction between the "revealing" of *aleitheia* (the ancient Greek notion of truth) and "revealing" as "challenging," which, he suggests, is suited better to describe the essence of modern technology.[18] For Heidegger, technology does not merely reveal Truth, but produces truths.

Therefore, by "challenging," Heidegger specifically speaks of violence to nature. As he puts it, challenging is "putting to nature the unreasonable demand that it supply energy that can be extracted and stored as such."[19] Though I am hesitant to make a claim that Heidegger was perhaps a crypto-environmentalist, it is not necessarily a stretch for one to understand the qualms he had with the concept of modern technology as involving the opposition between nature and humanity. Many interpreters of Heidegger consider him to be technophobic,[20] but I disagree. His act of revealing modern technology as a "danger" comes from a certain analytic response to industrial capitalism, a criticism of modern technology that moves from a materialist observation to a metaphysical argument. Modern technology moves to expedite production, or in his words, "unlock and expose" nature in the quickest way possible; therefore, modern technology allows for "maximum yield at minimum expense."[21] It converts nature to what he calls "standing-reserve," unconcealing energy concealed in nature and converting it for use in the forces of production in industry.[22] Heidegger here is not very far from Marx. The technologies that locked into place the industrial revolutions at the turn of the twentieth century not only alienate the human being from his labor and his "species-being," as Marx contends, but also, as Heidegger adds, throw the relations of humanity and truth onto a graph, mapping the most efficient cost-benefit ratio. The science (from the Latin root *scire*, which means "to know") of the Enlightenment, entailing a systematic accumulation of knowledge of nature in the name of human progress, very much relied on the technologies of production at the time.[23] Thus, Bruno Latour uses the term "technoscience" to refer to the technological and social milieu of science.

Consequently, Heidegger states, humans have limited agency against technology's ability to produce Being. "In truth," he says, ". . . precisely nowhere does man today any longer, encounter himself, i.e., his essence."[24] Yet, modern technology, an instrument of ordering of the real, *excludes* humans from this very process. Heidegger explains this paradox: "The coming to presence of technology gives man entry into That which, of himself, he can neither invent nor in anyway make," giving up the idea that "man, solely himself, is only man."[25] In modern technology, humans cannot be sure of their own humanity. Instead they are forcibly opened to the idea that humans are merely conglomerations of various social forces, opening up a vulnerable possibility that human knowledge and Being, even with technoscience, does not achieve mastery.

If one could make a criticism of Heidegger's notion of "revealing," it would be based on the Heidegger's seeming overdependence on the visual. After all, "revealing" and "enframing" are closely tied to an essay in which he describes the essence of modern technology as the formation of a "world-picture,"[26] a notion that the world could be explained as a totality to human consciousness. However, we must acknowledge that his use of the visual register in describing technology is a bit ironic. In fact, "revealing," as he uses it, does not at all mean the exposing of what is given, of the essence. Contrarily, it is more what he calls "emplacement" (*stellen*) or setting in order for the sake of representation or depiction. As Samuel Weber reminds us in his interpretation of Heidegger, technology can only depict something taking place by *putting* it in place, all in the name of securing the centrality of the human subject.[27] Hence, Weber highlights two major arguments advanced by Heidegger about the world-picture: (1) the claim to universality and (2) systematic tendency.[28]

Some would argue that the Internet does very much the same thing, playing a huge part in the increasingly visual ordering of the world, serving as a database of visual forms.[29] Google Earth, YouTube, and Flickr are just a few potential examples of this Heideggerian critique of technology. Undoubtedly, there is a degree of "pictorialization," as Weber would say, going on. However, to wholly place the function of the world-picture onto the Internet would be to fall prey to pronouncements in favor of neoliberalism and globalization that conveyed the McLuhanian message of the "global village," of a One World. As did alternative accounts of globalization, many theories of Internet aesthetics contest the characterization of the Net as just a proliferator of a Heideggerian world-picture.

Quite simply, the Internet exposes the tenuous bond between the visual and "the world." In this respect, digital design and new media art (including Net art) practices are disturbances in the enframing processes of the real. New media art on the Internet, "so replete with dazzling flashes of brilliance," similarly overloads the visual apparatus to "dizzying effects."[30] Visuality on the Internet has little to do with revealing or appearance. The natural world is an object that has come to be mediated to such an extent that it is doubtful that technology will have anything to do with gleaning more resources or information from it.

Anna Munster and Geert Lovink argue that Net art and new media art in general signal a shift from an information society tied to the computer to the "entwined fragmentation of techno-social networks."[31] New media, they argue, are distinctively distributed media with new types of aesthetics that do not correspond with singular forms or a singular medium.[32] They view the topology of contemporary media as uneven, which unavoidably necessitates a corresponding aesthetics. This observation can easily be used to frame the Internet. As I argue in the first chapter, the Internet and new media cannot be theorized in terms of singular gateway technologies. Again, to take the example of the Internet, many technologies are able to access it beyond the confines of the computer monitor, keyboard, and mouse.

Distributed aesthetics, according to Munster and Lovink, deal with situated, highly individuated media as allotted through both the "traditional" gateway of the personal computer and dispersed technomedia. They are characterized by asynchronous production as well as multiuser access to both material and informational artifacts, more suited to recent media practices such as art installations, e-learning, and so on. In other words, they are saying that media aesthetics are *not* located in their formal elements. Rather, they suggest that they *cannot* be found, dispersed in "the endless relaying of media."[33] A distributed aesthetics, Munster and Lovink argue, "might be better characterized as a continuous emergent project, situated somewhere between the drift *away* from coherent form and the drift of aesthetic *into* relations with new formations, including social (networked) formations."[34] In proposing such a framework for aesthetics of technomedia, Munster and Lovink are getting at what they perceive to be an oft–glossed over detail in media scholarship—that networks are theorized in a reductive manner. The reduction made by Munster and Lovink, which they call the "mapping information" model, treats media as merely the movement of information from server to network to client. However, as Kittler does, they insist that this model merely attempts to "tame the beast" of unstructured and floating info-bits in various techno-social networks.

Producing and Perceiving the Real

My friends Noah and Sam are currently living in the continent of Africa. Noah is working for an HIV/AIDS education through soccer program in Lusaka, Zambia, while Sam is researching Christian evangelicalism in Ibadan, Nigeria. It was not until recently that several of my friends decided to move abroad on a semi-permanent basis. To keep us, their friends, in touch with their travels and lives more generally, each has set up a blog using one of the more popular blogging host sites—Blogspot.com. These blogs, called travel blogs, are becoming more commonplace due to the increasing frequency and ease of air travel. As its name suggests, the travel blog is a blog that is administered and updated frequently by those who are traveling and is usually filled with stories from the road accompanied by digital photos. Though a decade ago, a blog would have been considered to require specialized knowledge and skill to maintain, today, any sort of traveling is occasion to start a blog. It is no coincidence that my friends Noah and Sam are first-time bloggers. I frequently read their blogs to see, quite literally, how they are doing.

By looking at travel blogs as one example, we can entertain the idea that visuality on the Internet has two contradictory functions. On one hand, images and videos on the Internet act as extensions of human perception as McLuhan once theorized. Indeed, in its status as meta-media, the Internet acts as the duct through which people exchange image and video files with others. On the other hand, visual forms play tricks on us all of the time. Digital cameras are no longer just the luxury of professional photographers. The phrase "Photoshopped" (in reference to editing software developed by Adobe) is used often when one sees images of celebrities in *Star* magazine or reports from the war in Iraq and more recently photos of bombings in Lebanon. All visual forms on the Internet are met with a degree of scrutiny, resulting from the insecure status of the "reality" of their content, the truth-value, so to speak. Though the Internet does, in many ways, heighten the visual sense of humans through the toppling of the laws of space and time, it also disturbs the phenomenological basis of "experience" upon which McLuhan's notion of human perception relies. The presentation of phenomena in technomedia is never a given. The Internet, in particular, is a hotbed for doctored images and video. One of the most popular websites for unusual material is eBaum's World (www.ebaumsworld.com), which logs more than a million hits per day. It is a website that plays host to irreverent video clips, pictures, prank phone calls, and cartoons. It features everything from Mike Tyson's famous bite to pitcher Randy Johnson's accidental murder of a bird flying by during a game. One very famous flash cartoon (an animated short movie made using the software called Flash, developed by Macromedia) is called "The End of the World," and parodies the Bush administration's War on Terror and the proliferation of nuclear weapons. Jon Stewart's *Daily Show* on Comedy Central does something similar. The show uses news stories from various television news sources to comically parody other "legitimate" cable news channels.

Websites such as YouTube and publications such as *The Onion* are havens for humorous, modified images of American political figures and the like (e.g., the popular image of President George W. Bush alongside several chimpanzees, which circulated around the Internet during the 2004 elections). Though usually these are not taken seriously, their popularity and amazingly fast diffusion among young people on the Internet indicates a capacity for finding pleasure in being misinformed or purposely tricked. Office humor in

America revolves around funny one-minute video clips, circulated through office e-mail listservs, which appeal to the most juvenile senses of humor. Yet, many cannot help but to click, watch, and LOL (laugh out loud). The Internet has the power to draw us close to a space filled with uncertainty, left without enough security of knowing that our senses are telling us the truth to catch our bearings.

While Munster and Lovink argue that the quick diffusion capability of the Internet marks a change in the aesthetics, Walter Benjamin, in "The Work of Art in the Age of Mechanical Reproduction (1936)," noted a similar change in perception with regard to cinema, which pioneered the previous centralized mode of distribution of media based on a production center (Hollywood) and various transmission peripheries (movie theaters). Perception changes as reproduction methods revolutionize. In other words, perception is altered by the changes in the logic and culture of capitalism. Benjamin's comments concerning mechanical reproduction suggest a "decay in aura," caused by the "desire of contemporary masses to bring things 'closer' spatially and humanly, which is just bent toward overcoming the uniqueness of every reality by accepting its reproduction."[35] The "original," in this new era of mechanical reproduction, he believed, was decreasing in value. He foresaw the daily scenes on Canal Street in Manhattan and the side streets of Bangkok where bootlegs and knock-offs are happily bought by consumers (I must admit that I once purchased a Folex). The *aura* of an object, its core, was no longer located in the originality or authenticity of the creator's work. Mechanical reproduction took away the object's history. The distance between the author of a work and its audience was, as Benjamin notes, "merely functional."

Walter Benjamin (1892–1940) was a thinker whose short life may have prevented us from benefiting from an enormously diverse body of work and a surely even greater body of work that would have been influenced by it. He wrote on drama, Jewish mysticism, Marxist theory, aesthetics, literature, and technology among other things. His friendship with and influence on fellow German scholar Theodor Adorno is responsible for his association with the Frankfurt School of Critical Theory. Though many of his Frankfurt School colleagues fled Nazi Germany successfully, he committed suicide on the French-Spanish border while on his own attempt to escape. He left unfinished *The Arcades Project*, his magnum opus detailing the street life of nineteenth-century Paris.

Aura: Benjamin speaks of *aura* as in decay in capitalism. In traditional, feudal society, objects had a certain aura about them, meaning that they were unique and singular. In capitalist societies, due to the onset of manufacturing and replication in the Fordist mode of production, the uniqueness of objects no longer existed. The craftsman or artisan gave way to the factory worker as the work of art gave way to commodities. Thus, the premodern romance of an artist laboriously painting at his easel is wiped clean by the image of the printing press spinning out thousands upon thousands of mechanically reproduced images.

Taking into account the decay of *aura* in the modern era, it is clear that *aura* might be forced into greater degrees of compromise in an era of postmechanical, postelectrical, now digital reproduction, indicating *another* shift in perception. Extending Benjamin's formulations from his famous essay, Samuel Weber introduces *mediaura* as a new theoretical context to understand the difference of "new media" from the old. He defines mediaura as "auratic flashes and shadows that are not just produced and reproduced by the media but which are themselves the media . . . a mass movement of collection and dispersion, banding together and disbanding."[36] *Mediaura* refers to a new kind of perception in an era of new media technology.

Mediaura describes the breakdown of perception's relation to consciousness. Psychoanalytic thought since Freud has argued that human faculties are not subordinate to the purpose and consciousness of the individual. Its most famous example, parapraxis, the Freudian slip, demonstrates the uncontrollability of speech, and dream life indicates that thought wrestles free from consciousness. In a similar vein, the Internet sways how individuals perceive the world. For one, it undermines the world-picture. On the contrary, the Internet presents the world in pieces. Therefore it is quite ironic that the Internet has spawned theories of global interconnectedness and network society when it provides several pictures of that society. It is home to what mainstream culture would deem bizarre, freakish, and downright immoral. As a platform to circulate alternative, dissenting, and sometimes wacky ideas and cultures, the Internet affects and structures the social unconscious by dislocating perception from consciousness and allowing for it to reside in the imaginary, placing knowledge on par with fantasy, making the social world, as we perceive it, freakish, bizarre, and generally scary.

The Internet does not do the work of covering over the absurd or securing knowledge for power and authority. Many argue that it does quite the opposite; it serves as a platform for the circulation of nonmainstream ideas. For some commentators, the rise in blog activity (as of 2006, 75,000 new blogs were being created daily)[37] indicates a new dynamic for information. Blogs are structured, much like the rest of the Web, in a system of links. Though initially claiming an underground status as web journals, blogs are now well-indexed and cross-linked thanks to sites such as Technorati and Del.icio.us, which serve as blog search engines.[38] Any blogger, or anyone familiar with blogs, can tell you that the hyperlink is of utmost importance for them. Links to news articles, other blogs, funny video clips, and so on make up the copy-and-paste culture of the "blogosphere." Blogs act as an alternative to mainstream media and, according to many media commentators, challenge their knowledge monopoly. In a recent BBC News article, Paul Reynolds highlights two important facets of blogs: (1) They are able to uncover things that the mainstream media cannot or will not and (2) they are a source for criticism of all kinds.[39] Opinion blogs are becoming a conventional means for users to obtain reviews about any type of commodity. News blogs are also contenders with major media sources for breaking news coverage. Some come to be extremely popular, like soldier blogs, which have won a large audience in the United States for their "on-the-ground" appeal. In addition, a blog called *Baghdad Burning*, written by Riverbend, the pseudonym of a young Iraqi woman, has come up for BBC Four's Samuel Johnson Prize for nonfiction, an award usually reserved for printed books.[40]

Recently, bloggers in China made international headlines for the attention they are receiving from their government. There are reports that the Chinese government has jailed independent bloggers for publishing "subversive material." Though Beijing has denied such wrongdoing, bloggers Li Zhi and Shi Tao are serving sentences for posting politically dissenting material on Internet blogs.[41] Furthermore, American companies in China have been reconsidering their operations procedures and exercising self-censorship in response to state pressure. The ensuing of such panic in reply to blogs makes the power and reach of blogs quite obvious.

All of this recent hype surrounding blogging highlights certain points concerning the Internet and knowledge. First, knowledge has yet to be freed from its authoritative usage even in the age of information. The pairing of knowledge and power that Foucault identified remains a truism on the Internet. China's actions are just one example of the regulatory tendency of nation-states, reacting to the crises in cogency of nation-states worldwide. However, the Internet, though not a full-blown liberating force, contests the power effects of the regulatory schemes of disciplinary societies. Some may complain that everything on the computer screen must be met with a high degree of scrutiny because of the very trickiness of "information" on the Internet. This, I am suggesting, is not necessarily a bad thing. The Internet strips away the aura of "neutrality" of past media and technologies, making it clear that knowledge is constitutive of its mediating technology. The Internet is the space where that which was swept under the rug in modernity peers its head out, calling attention to itself. Those of us who exist in the folds of the Web are forced to watch, listen, and, in some ways, be seduced (think how much porn- or sex-related junk mail one receives in any given week). It is not a "virtually real" space; it is simply a virtual space, with its own behaviors, mechanisms, and idiosyncrasies. On the Internet, we encounter that which cannot be "real," yet our drives lead us to believe in it anyway.

Notes

1. Friedrich A. Kittler, "On the Implementation of Knowledge: Toward a Theory of Hardware," at Peter Krapp's theory website the Hydra, www.hydra.umn.edu/kittler/implement.html.

2. Roland Barthes, *Mythologies* (Paris: Editions du Seuil, 1957), 116.

3. Barthes, *Mythologies*, 116.

4. Anson Rabinbach, *The Human Motor: Energy, Fatigue, and the Origins of Modernity* (Berkeley: University of California Press, 1992), 45.

5. Rabinbach, *Human Motor*, 46.

6. John B. Thompson, *The Media and Modernity: A Social Theory of the Media* (Cambridge, UK: Polity Press, 1995), 3. Emphasis mine.

7. Jurgen Habermas, "The Public Sphere: An Encyclopedia Article," in *Media and Cultural Studies: KeyWorks*, ed. Meenakshi Gigi Durham and Douglas Kellner, rev. ed. of *KeyWorks in Cultural Studies* (Malden, Mass.: Blackwell, 2006), 73.

8. Habermas, "Public Sphere," 75.

9. Habermas, "Public Sphere," 77.

10. Michel Foucault, *The Order of Things: An Archaeology of the Human Sciences* (London: Tavistock, 1970).

11. See Marshall McLuhan, *The Gutenberg Galaxy: The Making of Typographic Man* (Toronto; London: University of Toronto Press, 1962).

12. Marshall McLuhan, *Understanding Media: The Extensions of Man*. (New York: McGraw-Hill, 1964), 20.

13. McLuhan, *Understanding Media*, 20.

14. McLuhan, *Understanding Media*, 21.

15. Martin Heidegger, *The Question Concerning Technology, and Other Essays*, 1st ed. (New York: Harper & Row, 1977), 4.

16. Heidegger, *Question Concerning*, 5.

17. Heidegger, *Question Concerning*, 8.

18. Heidegger, *Question Concerning*, 13.

19. Heidegger, *Question Concerning*, 14.

20. For a reading of "The Question Concerning Technology" that does not fall into the easy trap of calling Heidegger a technophobe, see Mark Poster, *What's the Matter with the Internet?* (Minneapolis: University of Minnesota Press, 2001) and Bernard Stiegler, *Technics and Time 1: The Fault of Epimetheus* (Palo Alto: Stanford University Press, 1998).

21. Heidegger, *Question Concerning*, 15.

22. Heidegger, *Question Concerning*, 15.

23. See Stanley Aronowitz, *Science as Power* (Minneapolis: University of Minnesota Press, 1988) and Patricia Ticineto Clough, *Autoaffection* (Minneapolis: University of Minnesota Press, 2000) for two different approaches to the relation between knowledge-production and technology.

24. Heidegger, *Question Concerning*, 23.

25. Heidegger, *Question Concerning*, 31.

26. The essay to which I refer is "Time of the World-Picture" in *The Question Concerning Technology, and Other Essays*, 1st ed. (New York: Harper & Row, 1977).

27. Samuel Weber and Alan Cholodenko, *Mass Mediauras: Form, Technics, Media* (Stanford, Calif.: Stanford University Press, 1996), 79–81.

28. Weber, *Mediauras*, 80.

29. Weber, *Mediauras*, 79.

30. Weber, *Mediauras*, 318.

31. Anna Munster and Geert Lovink, "Theses on Distributed Aesthetics. Or, What a Network Is Not," in *fibreculture* no. 7 (2005), at journal.fibreculture.org/issue7/issue7_munster_lovink.html.

32. Munster and Lovink specifically point to Lev Manovich's *The Language of New Media* (Cambridge, Mass.: MIT Press, 2002) as

an example of media formalism that does not take into account the distributed nature of media today.

33. Munster and Lovink, "Distributed Aesthetics."

34. Munster and Lovink, "Distributed Aesthetics." Emphasis in original.

35. Walter Benjamin, "The Work of Art in the Age of Mechanical Reproduction," 1935/1936, at www.jahsonic.com/WAAMR.html.

36. Weber, *Mediauras*, 106.

37. Statistic from Michael Geist, "Locking Down Our Digital Future," *BBC News Online*, February 8, 2006, at news.bbc.co.uk/2/hi/technology/4690188.stm.

38. Bill Thompson, "Can We Stop the Bloggers?" *BBC News Online*, February 24, 2006, at news.bbc.co.uk/2/hi/technology/4747818.stm.

39. Paul Reynolds, "Bloggers: An Army of Irregulars," *BBC News Online*, February 9, 2006, at news.bbc.co.uk/2/hi/in_depth/4696668.stm.

40. "Blogger up for Non-fiction Award," *BBC News Online*, March 27, 2006, at news.bbc.co.uk/2/hi/entertainment/4847424.stm.

41. "China Defends Internet Regulation," *BBC News Online*, February 15, 2006, at news.bbc.co.uk/2/hi/asia-pacific/4715044.stm.

3

Space, Time, and Matter in the Virtual

> "The technical artifacts that surround us are more than just extensions of ourselves; they shape and mediate our experience of life and the taking place of space and time."
>
> —Belinda Barnet[1]

On May 5, 2005, Professor Belinda Barnet found herself in Morocco. Like many of us who are able to travel abroad to lands foreign to us, she was equipped with a guide. In this instance, her guide was not a book she picked up from an airport newsstand or the locally placed but internationally owned chain bookstore. To get around, she was using her mobile phone, equipped with GPRS (General Packet Radio Service).[2] Though she was surrounded by smells of foods, words of languages, and walls of buildings strange to her, she was not lost. Her mobile phone informed her of all of the goings-on around her in her own language. Though having traveled a long way from her native Australia, Barnet was not a complete stranger. Yet she was not a local either. As she explains, she was somewhere in between:

> I bring information with me to this place, and this information *mediates my experience*; the territory surrounding me serves as a surface of projection for data. . . . Every street building has a layer of virtual graffiti I can summon in an instant, my experiences will in turn be archived and will form part of this collective inscription. . . . Using my device, I retrieve 243 entries for Café Toubkal on the East side of the square; if I squint I can see it through the jumble of stalls. . . . As my feet slide upon thousand-year-old stone, I am at once traveling through networks and central servers back in Australia, my details handed on via invisible network handshakes across the globe, my trajectory recorded. I am not lost, I am identifiable; I am a string of information events.[3]

Barnet's description of her experience in Marrakech tells us quite a few things about the nature of technomedia today with respect to the experience of space and time. Drawing from Bernard Stiegler, Barnet suggests that technomedia today bring together three spheres once thought to be mutually exclusive—"the industrial technical system, globalization, and mnemotechnical systems"—to form a "global mnemotechnical system" that "incorporates digital information networks like the Internet as well as the real-time information events of individuals."[4] Therefore, in her case, the experience of space and time was not simply determined by hardware but was "an 'interweaving' of this global system with real space."[5] Thus, Barnet's experience of Morocco is not solely a product of her own empirical observations. Her experience of space/time is not a simple aggregate of perceptions limited to human senses or even her "cultural bias" as an Aussie, but is a blend of those with the mediating information of her mobile phone.

Spaces and Times Past

If we take Barnet's account of her technomediated experience of space and time as commonplace, some theoretical backtracking is in order to see how it is that we have come to the point where humans are "info-mobile," to use her words. Prior to the rise of technomedia, in spite of the various advances in the theory of relativity by Einstein and Heisenberg, space and time were, for all intents and purposes, *fixed* entities. Undoubtedly, Einstein's theory brought forth a powerful challenge to linearity, the dominant interpretation of time, in physics and philosophy, where the seeds for their modern definitions were laid. But in the social world, not many could see or even feel the implications of Einstein's theory.

Zygmunt Bauman suggests that time, in fact, begins with modernity. All notions of space and time prior to modernity are, as he describes it, a *history* of time. Explaining the modern notions of space and time, Bauman writes, "'Space' is what you can pass in a given time, while 'time' is what you need to pass it."[6] In other words, the interrelatedness of space and time was directly correlated with the capacity of the body, the human muscles. He goes on to clarify that these definitions of space and time were founded strictly on *human* measures. Though some could obviously run and walk faster than others, "the replacement of one pair [of legs] with another would not make

Zygmunt Bauman (1925–) is professor emeritus of sociology at the University of Leeds. Recently, Bauman has been seen as a postmodern thinker, but his earlier works vary in theme from socialism to the Holocaust. Originally from Poland, Bauman left his home country in 1968 when the government drove out the remaining Jews within its borders. Bauman, along with many other intellectuals, lost his university post and fled. After a brief stint in Israel, Bauman eventually settled in the UK and remains there today.

Since his retirement in 1990, he has published an astonishing number of books (over twenty). Some of his key works include *Modernity and Ambivalence* (1991), *Postmodern Ethics* (1993), and *Liquid Modernity* (2000), which contains his most succinct statement on issues of space and time.

large enough difference to call for measures other than the capacity of human muscles."[7] However, with modern industry and technology, the need for "deployment of something else than human or animal power" was met and space/time as a functional derivative of the physical body was eclipsed.[8]

According to Bauman, modernity delinked and "distanced," to use the words of Anthony Giddens, space and time from one another. The philosophies of Newton (his work on the physics of acceleration and distance in particular) and Kant (his separation of space and time as two universal and mutually exclusive categories) expressed this view very clearly. In modern times, with industrialization, new modes of transport such as the automobile redefined those old definitions of space and time. Time was no longer

Anthony Giddens (1938–) is one of the most prominent British intellectuals in the world today. He is the author of over thirty books and has taught at Leicester and Cambridge universities. He served as director of the London School of Economics and sits at the House of Lords for the Labour Party, a political appointment given by Tony Blair, whom he advises. Though today Giddens is seen as the architect of the Third Way policies of the Blair government, he was influential in social theory for his writings on structuration theory, modernity, identity, intimacy, and, most cogent to our present discussion, "**time-space distanciation**." Time-space distanciation is Giddens's way of explaining the way in which modernity "disembeds" and "reembeds" locales and context, stretching identities, social relations, and societies beyond local contexts. Thus, in more modern societies we find the mobility of identities and social interactions beyond local communities. This is in large part due to the technological innovations that came out of the evolution of industrial capitalism. See chapter 1 for more about modernity and technology.

related to distance in the same way because of its contingency upon new means of transport—the hardware. Thus, Bauman concludes that "the time needed to travel ceased to be the feature of distance."[9] Modernity was the age of acceleration, he rightfully claims.

Modern technology—hardware—differentiated space from time because it was able to directly manipulate time to produce different temporalities. Bauman refers to this era of hardware as "heavy modernity." However, in today's "light" or "liquid modernity," characterized by what he calls "software capitalism," the delinking of space and time must be reconsidered. Technomedia, Bauman argues, annihilated space, *not* time.[10] "Instantaneity," a key feature of liquid modernity, "denotes the absence of time as a factor of [an] event."[11] With that said, Bauman asks whether time, too, is on the road to its own dissolution.

For Bauman, modernity's rationalizing logic of mechanical acceleration, exemplified by automobiles and industrial manufacturing, acted as a harbinger for liquid modernity's "seductive lightness of Being."[12] The logic of software capitalism, based on "near-instantaneity," demonstrated a new form of value. On one hand, time in "heavy modernity" was harnessed and rationalized into the work day to maximize surplus value, as Marx lays out at the beginning of *Capital I*. On the other hand, time in "software modernity . . . as a means of value-attainment tends to approach infinity."[13] According to Bauman, liquid modernity brings to the fore "the new irrelevance of space, masquerading as the annihilation of time."[14]

Bauman's insistence that time's apparent dissolution is due to the annihilation of space remains firmly planted in older ideas of materialism. Though distance as the defining characteristic of space has indeed given way, a new logic, one that has to do with the concept of matter, has emerged along with media technologies that not only transport material entities further and quicker, as Bauman argued, but also challenge prior notions of "the material." With changing notions of space come changing ideas of what is able to occupy space—namely, what counts as "matter." Bauman overlooks the new modes of thinking about matter that have resulted from the effect of media technologies on human life. Thus, technomedia call into question the strict formulations of physical space and

Jacques Derrida (1930–2004) was an Algerian-born French philosopher and director of studies at the École des Hautes Études en Sciences Sociales until his death in 2004. He was one of the most influential French academics in North America, due in part to his frequent speaking engagements and visiting professorships at various institutions. His prolific career included numerous books on art, architecture, literary theory, linguistics, and, closer to the end of his life, politics and ethics. Though he did enjoy great popularity and intellectual allegiances in the United States, to many laypeople and some academics, he was responsible for nihilistic and apolitical views such as "deconstruction" and "postmodernism." Despite this largely misunderstood charge, Derrida remains an important figure in various fields of study. Amid numerous new concepts that he was responsible for introducing to the world, there is one in particular that bears the most significance in this book.

Subjectile: *Subjectile* is a term that Derrida takes from French playwright Antonin Artaud. As interpreted by Derrida, the subjectile is an attempt to rethink the problematic of the human "subject" that has been at the center of many intellectual discussions around poststructuralism. The discourse on the "death of man" (Foucault) and the "death of the author" (Barthes) initiated yet another layer of discourse that signaled the "death of the subject," bringing to light the privileging of human subjects, an anthropocentrism or homocentrism haunting the human sciences. The subjectile, as discussed in the chapter on subjectivity, presents an entrée into what I believe to be grounds for new thinking about "users" of computer technologies beyond subjectivity and toward lurking, which, much like Artaud's subjectile, opens up to new ideas regarding the connectivity between media technologies and humans. The concept is discussed at length in chapter 4 on subjectivity.

linear time, opening up to dynamic articulations of spatialities and temporalities. In technomediated worlds, space is not a predetermined demarcation of physical territoriality, but a dynamic process of spatial formations; time is not a transcendental, linear movement of human history, but a set of mediated temporalities.

Communicative Events and Contexts

The most easily understood example of the vast influence of technomedia on notions of space and time is in telecommunications. Following Bauman, the annihilation of space in technomedia complicates the notion of communication. What is the context of a conversation occurring over text messages or instant messages? Who is present and who is not? Is there even a "presence" of which to speak? Where is this conversation taking place? These spatial and temporal elements are not as certain as they once may have been in times past when communication meant strictly face-to-face interaction. Yet as Jacques Derrida reminds us, even during times when "telecommunications" meant letter-writing, the notion of context as a fixed moment and/or place in time and space was precarious, if not wholly misguided. He writes:

> It seems to go without saying that the field equivocality covered by the word *communication* permits itself to be reduced massively by the limits of what is called a *context* . . . [b]ut are the prerequisites of a context ever absolutely determinable? . . . Is there a rigorous and scientific concept of the *context*?[15]

Derrida suggests here that communication is predicated on an idea of context as a fixed set of rules or boundaries in which a communicative process occurs. He notes that writing is afforded a homogeneous space of communication, where the consciousnesses of both reader and writer are understood to be, to use Derrida's words, *extended*.[16] Quite obviously, he is wary of the assumptions made when context refers to an equal field of communication. He asks whether the fixed set of rules or codes of a context could ever be identified. Though he does not dispute the effect of such rules or codes, Derrida suggests that they have a mobility or flexibility about them. He notes that communication (in the traditional sense)—a transmission of unitary meaning—does not ever really occur without an *absence* of some sort.[17] The fixed presence of bodies needed in face-to-face communications is not needed, for example, in letter-writing. The structured physical absence of both reader and writer that Derrida highlights in regards to written communication can be imagined to be far more pronounced in an era where the context of communication is ever-changing due to the mobility of technomedia.

Text messaging is one phenomenon that illustrates technomedia's effect on context. Derrida points to the signature as that which represents the actual or empirical nonpresence of the signer, leaving a trace of a communication of some sort, as unstable as it might have been.[18] Similarly, a text message stored in the inbox of my mobile phone from someone in my contact list lacks a context:

Received: 12:14PM
From: Sara
Message: Yes, c u in 5.

What can one deduce from this without context? I do know that a communication did occur, yet I have not a clue what Sara may have had in mind in this message. Where were we meeting? If we were meeting, what was it for? Where was it? Was this message specifically meant for me or was it a mass text message? All of these questions, in essence, of one's context or, better put, one's location in space and time, are uncertain as technomedia's mobility and speed leaves us with nothing but handles, screen names, and phonebook entries—postmodern signatures.

According to cultural theorist Paul Virilio, the coupling of increased mobility and speed in contemporary technomedia signals "the defeat of the world as a field, as distance, as matter."[19] In his analysis of current trends in war strategy, Virilio points out that the "strategic value of the non-place of speed has supplanted that of place."[20] The change in war strategy, Virilio indicates, results from a change in the technologies of the military/war. The most advanced military technologies allow for the launch and detonation of weapons to occur thousands of miles away from the target. This type of warfare differs quite a bit even from World War II, in which military leaders, while holding the most powerful of military tactical innovations in the form of the atomic bomb, still needed to send bomber planes to the Pacific Theater to commit acts of atrocity. In other words, physical intimacy in traditional and modern warfare has been eclipsed by innovations in war technologies. Today, nuclear weapons can be launched to targets thousands of miles away, as the recent nuclear tests of North Korean Taepodong missiles show. Though many Western nations have frowned upon the nuclear capabilities of Iran and North Korea, the United States, which has

Paul Virilio (1932–) is a French urbanist and theorist of technology who studied with phenomenologist Maurice Merleau-Ponty at the Sorbonne after working with Matisse. With regard to his self-described "marginal" view of technology, his work has dealt with three major areas—war, speed, and perception. He argues that military strategy and technologies determine the shapes of cities and societies in general, resulting in what he calls a "war model." He and many other recent theorists like Giorgio Agamben and Jean Baudrillard have turned their attention to the rather unending nature of war, especially in a post-9/11 global context. Unlike Agamben, who recently has been the most oft-cited of theorists of war, Virilio has spoken about perception and speed, focusing particularly on the movement of the images of war in technomedia.

Dromology: Virilio introduced *dromology* in *Speed and Politics* (1977) in order to shed light on the importance of the speed of the military-industrial complex in affecting changes in society. It is in fact a social theory based on investigating the role of speed and acceleration in history by looking at its functions in everyday life. Virilio argues that the speed of an event affects the very nature of all entities involved. Additionally, he suggests that technology also affects human perception and experience by rendering them fragmented and discontinuous. For our discussion, Virilio treats the changes in perception and experience as driven by alterations in time. Thus, the Internet serves as a great example of the instruments that Virilio argues are driving social change.

done much of the objecting, particularly during its soldier-recruitment campaigns, flaunts its own technological sophistication.

Regarding the technological undervaluation of space, Virilio argues that military action has been "miniaturized" and that the "time of the finite world is coming to an end,"[21] as the world continues to resemble a game of Battleship or Risk. War is just one indication of what Virilio calls the "generalized interaction" that occurs as a result of globalization. Globalization has made clear the unity of all actions, and whether miniaturized or generalized, they now have become far-reaching in their effects. The scale of military action and, Virilio is quick to note, military media coverage (e.g., embedded reporters and helmet-cam) has changed, undoubtedly because of the change in space and time. The "instantaneity and ubiquity" of global interaction reside most evidently in military technologies that have helped globalize the world.

Microtempos

In the theory of relativity, Einstein argues that neither space nor time is a fixed constant. Einstein's critique comes, in part, through an exposition of the methodological assumptions in Newtonian physics. Among various examples, one of the most striking is that by which Einstein critiques the instrument of measurement itself—the clock. Unconvinced that "clocks go *at the same rate* if they are of identical construction,"[22] he asserts that clocks behave differently according to their spatial arrangements. In one example, he explains that two clocks, one in the center of a rotating plane circular disc (reference point K) and the other placed at another point along the plane (K^1), due to the effect of inertia, would operate at different rates. Thus, when observed, the clocks show different times. Without getting into the mathematical ins and outs of his argument, Einstein's point is that "it is not possible to obtain a reasonable definition of time with the aid of clocks which are arranged at rest with respect to the body of reference."[23]

Today, a century after Einstein's initial articulation of the theory of relativity, there are other temporalities that are challenging the clock-time of modernity. With increased mediatization, these modern tenets of existence (space and time) have been questioned and changed without much notice. Perhaps this is so because the virtualities of technoscience are now *lived*, not merely thought. As Einstein suggested, the clock, coupled with the calendar, did not simply "measure" time, but generated times. Against Bauman, Einstein's theory forces us to consider the possibility of multiple times. The insidious technomedia of the present—the Internet, television, and mobile phones—overhaul the security of linear time as the dominant mode of temporality, even more so than the theoretical physics of Einstein and those who came after him. In effect, the concept of time could be replaced with something more along the lines of micro-temporalities.

Televisual Time

In *Still Life in Real Time*, Richard Dienst argues that televisual time has not only undercut clock-time, as Einstein's theory did, but also produces alternative temporalities. "The dynamics of causality, sequence, and logical order," he writes, "are subjected to the demands and exigencies of more abstract [televisual] temporalities."[24] He is pointing to television, and I think other technomedia as well, as a temporal force equal in significance to clock- and calendrical-time. Televisual time, for Dienst, bifurcates most often into what he deems *still time* and *automatic time,* determined mostly by television's visual quality—the images on-screen.[25] He defines still time as "a series of pure instants without any relation" or, as I see it, a slideshow effect. Whereas still time refers to a break in flow, automatic time is "an *extension* of movement by another kind of mechanical movement,"[26] resembling a stutter. It

> appears when an image is switched on and left running, so that it is no longer an image of *something*: it is the time of the camera's relentless stare, persisting beyond the movements of objects and scenery that pass before it . . . automatic time opens onto an anticipated future: it is an image waiting for its event to happen.[27]

Dienst here is referring to something along the lines of what, in technical terms, is called a pan-shot, where the camera turns horizontally on a vertical axis. Television modulates between automatic and still quite nicely. As Dienst says, "[it] allows for the instant to disrupt the continuous and vice versa."[28]

Furthermore, according to Dienst, on top of the constriction and extension of still and automatic time, the conventional use of the remote control to "zap" between and across channels also affects the experience of temporality on television. "Zapping" elaborates the relationship between flow of visual images and temporality. Along with the "mixes" or "switches" of television (i.e., commercial breaks), changing the channel opens up the viewer to

Richard Dienst (1962–) is associate professor of English at Rutgers, the State University of New Jersey. He is a cultural theorist who specializes in television and film. His *Still Life in Real Time* is a significant contribution to television studies, in which Dienst offers readings of several key theorists of television and technology. It is interesting for his interpretations of Deleuze's concept of "time-image" as well as Marx's labor theory of value. In addition, *Still Life* gives a highly original account of "zapping," the practice of switching channels with the remote control. Dienst notes that zapping is another plane of temporality, even beyond the greater "flow" of television controlled by the centralized broadcast. With the ability to zap, the viewer is able to cut through the dominant temporality produced by video—the sequence of moving images.

"innumerable points of visibility" and "streams of images . . . which nevertheless share the same moment and which always stand ready to emerge into a new present."[29] Televisual time displaces the traditional tri-partite (past, present, future) understanding of time for an "always present tense."[30] For Dienst, whether our sets are turned on or off, whether we are paying attention or not, the television's visuality presents presence constantly, without necessarily fixing our physical bodies—what is aptly called *telepresence*.

Virtual Temporalities

The work of Manuel De Landa introduces us to the concept of the virtual, which has implications for our present discussions of time and space. In "Deleuze, Diagrams, and the Open-Ended Becoming," he provides a flat-out rejection of causality, the temporal mode on which linear time was grounded. He aims to rethink time as not happening *to* us but *by* us and even by "nonhuman agencies."[31] This radical view of time does not fall into the essentialism of the Enlightenment thinkers—Newton, Descartes, and Euclid, in particular. De Landa presents Deleuze's critique of nineteenth-century physics to indicate the possibility of what he calls "an open-ended future." He argues that classical physics has essentialized and predetermined the past and the future. Classical physics, by its rationalist and mathematical means, made the "past a given and the future closed."[32] In a limited fashion, nineteenth-century physics viewed the world as "a fixed set of beings to which our theories correspond like a reflection or snapshot."[33] "Clockwork determinism," De Landa suggests, made the world timeless, for time was not given any creative role beyond that which one observes change against. This position, known as causality, created a vision of the world in which events occurring within it could be explained (thus caused) by earlier events.

The philosopher of science Ian Hacking understands *causalism*, the advocacy of the idea of causality, as follows: "We must understand why a certain type of event regularly produces an effect. Perhaps the clearest proof of such understanding is that we can actually use events of one kind to produce events of another kind."[34] Though not speaking in his own voice, Hacking explains that causality rests upon the ability to repetitively demonstrate correlations between two events. Causality, indeed, is not so much a cause and effect, but a relation, and a decidedly rigid one.

To counter such a premise, De Landa starts with a rethinking of matter as "intensive," with its own resources for "generation of form from within,"[35] what he refers to as *morphogenesis*. Unlike causality, this intensive process does not include "change over time" (modern science's favorite phrase indicating causality), but instead requires, in the words of Henri Bergson, a "divergent actualization." Here, De Landa, mediated through Deleuze, theorizes matter as not simply being acted upon by external structures, but capable of intensive or inner processes. In other words, matter is able to create, and operate in, its own time. In much the same way, Immanuel Wallerstein, incorporating the ideas of French historian Fernand Braudel and chemist Ilya Prigogine, argues for a "plurality of social times," multiple times interwoven into "a sort of dialectic of durations."[36] Times are always under negotiation between the microscopic event and an infinite eternal reality, what Wallerstein calls the "unexcluded middle."[37] The object of critique, for both De Landa and Wallerstein, is Newtonian physics' obsession with equilibrium. Wallerstein, taking from Prigogine's concept of "deterministic chaos," and De Landa, reworking Deleuze's conception of matter, think of time beyond its usual linear, beginning/end pairing.

De Landa uses the example of the feedback loop as a nonlinear relationship that is indicative of his Deleuzian formulation of matter.[38] Feedback is a circular causality—effects reacting back on their causes in a dynamic fashion. This is precisely how microphone feedback works: The microphone picks up the sound of itself amplified in a speaker. In a feedback loop, beginning and end become indiscernible (resulting in the familiar sharp howling noise that occurs at music venues); the sound loops and continually emits unless the

Manuel De Landa (1952–) is adjunct associate professor at the Graduate School of Architecture, Planning, and Preservation at Columbia University. Born in Mexico and based in New York City, De Landa is one of the most radical thinkers today, meshing together his specialized knowledge of complexity and nonlinear theory with Deleuzian philosophy and science. His recent *New Philosophy of Society* (2006) is a clear introduction to the various aspects of his thinking. For the discussion of space that takes place in this book, De Landa's exposition of *morphogenesis* is of import.

Morphogenesis: Quite simply, *morphogenesis* is an oxymoron. Genesis, as traditionally thought of, has very little to do with morphing or change. The traditional theological definition of genesis, that of Creation, regards the ending already determined at the beginning. However, De Landa argues that this concept of genesis is not only linear, but also holistically accepting of the power of external forces upon all matter. Instead, De Landa introduces "morpho" to connote a becoming, an ontology that is not determined by external forces. In biology, morphogenesis has a very specific meaning in relation to cell growth and cell differentiation. Using this as a point of departure, De Landa offers a redefinition of "matter" as not simply inert objects, but assemblages of dynamic forces. Thus, matter has the potential to self-organize *intensively* as opposed to extensively.

chain is broken (usually by moving the microphone away from the speaker). Without a marked beginning or end, linear causality is destabilized and gives way to something more closely resembling repetition.[39] In this way, the future, undetermined by rationalism, is open-ended for De Landa.

Although time for De Landa, Dienst, and Wallerstein is malleable and multidirectional, the figure of Father Time (as well as Mother Nature for that matter) is a trope that is not easily shed, even in realities immersed in media technology. For modernity to have faced head-on the implications of Einstein's denaturing of time, a difficult fissure would have had to result, one which would have led us to examine the theological roots of Western metaphysics.

If we apply Euclidean geometry to the Book of Genesis, time has a clear beginning, a source, an original point (e.g., [0,0] on a four-quadrant graph). The past, therefore, could be drawn as the line between the origin and the present (X,Y), and the future everything beyond the present. This is largely the underlying assumption of the concept of time in Western metaphysics. We get the phrase "point in time" from this type of thinking. Its roots are in the creation narrative, as God spoke the universe into existence. However, the classical theologian St. Augustine, in his *Confessions*, questions the relationship between time and God, using specifically the story of Genesis. Repeatedly, Augustine asks whether God exists within time. His answer is no; God, in the beginning, created time along with the earth, heaven, sun, and moon. Augustine concludes that time is for humans, while God exists outside of time. Time is of our human universe, not God's. Hence, Augustine concludes that *time is not autonomous, but simply a means of measuring events utilized by human memory, both individual and collective, as opposed to a constant divine (or natural) force.*

If technology, memory, and science each has its own time, human existence as unique and special comes to a predicament. Its purported superiority and sanctity, proven through scientific knowledge, begins to crumble under the weight of the present as technomedia threaten the uniformity of history and the predestination of the future. Time rethought in this way imparts an ever-present time—what Augustine called eternity and what the philosopher Emmanuel Lévinas deemed nothingness. We are plugged into multiple times: Our own virtual times influenced by the various technomedia that make up our lives. However, there is no such thing as "technomediated time." Media technologies constitute various temporalities that, by design, are not constant. Rather, they are dynamic; that is to say, they take hold for specific tasks and events only. The steady meter of modern life is superseded by the staggering and disruptive microtempos of postmodern life. Technology and media have contributed to a stripping away of the blanket of security under which humans were able to be sure of, if anything, time itself and along with it their own history and future. With neither history nor future, existence is ripped from the existent and Being becomes filled with emptiness. In this moment, existence may resemble insomnia, as Lévinas analogized, a never-ending consciousness, a restless vigilance to oneself, where time begins nowhere.[40] Without such a blanket, our vulnerabilities are left open for all (including ourselves) to see, for better or for worse.

Multiplying

As with time, space on the Internet, too, has multiplied and dispersed immeasurably; the Internet produces not only times, but also spaces. Though some may say that the necessity of servers and hard disk space for the computerized exchange supports the claim that centralized, physical space remains extant, which Bauman rightly notes, I would argue along the lines of Manuel De Landa that different, more dynamic types of spaces also now exist. We are at a point where the virtualization of space is no longer limited to fantasy, as the boundaries between fantasy and "empirical" reality are challenged in a very powerful way by the ubiquitous presence of the Internet and other technomedia.

If we are to discuss in particular hard drive space, or memory, as it is called, we can trace the overall decentralizing trend that the concept of space has undergone. Hard disk space is mainly utilized for information storage. Due in large part to the language of digital media coming from the 1970s, the popularized lexicon of popular operating systems like Mac OS X and Windows (e.g., "files," "folders," "storage") was very much based on a so-called material and physical ideal of space, taking from previous media and technologies, despite the fact that the very novelty of digitalization came from its ability to store information with a truncated amount of space (nearly none) necessary. In the example of word-processing—one of the earlier functions of the personal computer—the information of a hundred-page manuscript could be stored on, at first, an 8-inch, and eventually a 3.5-inch diskette. The size difference is significant and, in many ways, fascinating. In today's world, storage continues to vanish into ethereality. The introduction and subsequent incorporation of new portable storage devices called "flash drives" has brought to the forefront the potential for multiple spaces for a single document.

A hundred-page document today can exist on several storage devices. It could be on the personal computer on which it was first written. It could be on a USB–flash drive

Flash drives are data storage devices that have superseded floppy disks. They often utilize the Universal Serial Bus (USB). As its name suggests, the flash drive makes use of flash memory, a type of memory that is nonvolatile and reprogrammable. Nonvolatile simply means that the device does not need power; reprogrammable refers to the erasability of the data placed on the device. The uses of the flash drives are not limited to data storage; they can also be used as mp3 players so that a user may simply drag-and-drop music files to play back later on.

on which the author may have saved it in order to take it to work or school. Additionally, it could also exist on the computer at work or school from which she could have done some more work on the document. Or, perhaps, she could have avoided physicality altogether and e-mailed it to herself, in which case the document would also exist on a huge server of her e-mail account provider. Therefore, the original, in many ways, could be difficult to trace in this space of spaces. With the ability to modify the contents of any document on virtually any personal computer, a work no longer becomes singular, but plural, as Jean-Luc Nancy puts it, or multiple.

Additionally, it should also be noted that webmail, e-mail accounts on the web, is increasingly growing with respect to storage space. For example, Gmail, Google's version of this, offers almost unlimited storage. Though computer word-processing was seen as merely an extension of Gutenberg's print technology (since printing documents was once the main function of the personal computer), we can now see that word-processing in many cases takes a backseat to other functions on the computer. Yet, looking at word-processing and its ramifications on concepts of space and storage, we can see the withering of the separation between the empirical real and the virtual.

Subsequently, we can no longer see space as separate from technomedia. Space is wholly connected to the technologies that constitute it and that it constitutes. Spaces are no longer enclosures in the ways that buildings, neighborhoods, cities, states, nations, and geographic continents were. It is difficult to say with precision and accuracy where one is, especially on the Web. If someone is flying from San Francisco to Seoul and telephoning to New York, where is she? She is everywhere and nowhere; her location is on the brink of not even mattering anymore. To take this point into virtual space, if someone is browsing the Web, with multiple windows (or tabs) open to different websites, is there a way to track this person's location? Where was Belinda Barnet—Morocco or somewhere else?

Certainly, there are surveillance methods that use IP (Internet Protocol) addresses to track users of the Internet and so forth. In response to tracking strategies such as these, programmers have developed antitracking software such as xB Browser, an executable file that facilitates anonymous browsing.[41] There is space, but no longer location. Whatever way space is constituted, defined, or restricted, the Internet, usually at the moment of definition, connects and disconnects the space into another context, or, if you will, another space or spaces. The Internet denies positionality and self-location traditionally rooted on fixedness.

With that said, the "virtual," as De Landa suggests, may be said to have superseded "reality." In arguing for a new, nonlinear, and dynamic thinking of space, he conceives of a heterogeneous space consisting of a population of multiplicities, each of which is a topological space of spaces, with the capacity for progressive differentiation. The virtual continuum, he says, is a space of spaces.[42] Though the word "virtual" has been around to describe the Internet and technomedia in general by many thinkers, De Landa may even be suggesting the breakdown of the separation between the real and the virtual on an ontological level, as Patricia Ticineto Clough has argued.[43] De Landa and Clough do not use "virtual" as an adjective—as in "virtual" reality—in order to suggest that the Internet is a representative space or simply another version of reality. In fact, the usage of "virtual" by De Landa and Clough puts them in direct opposition with those like Howard Rheingold, who is fanatical about the *representative* possibilities of the Internet in particular.[44]

The concept of space that De Landa and Clough maintain brings us to the point that information is no longer the

Patricia Ticineto Clough (1945–) is professor of sociology at Queens College and the Graduate Center, City University of New York. She has written extensively in the fields of social theory, feminist thought, social science methodology, and technologies. More recently, her focus has turned to the interstices of affect, value, and biotechnologies. Her books include *The End(s) of Ethnography* (1992), *Feminist Thought* (1994), *Autoaffection: Unconscious Thought in the Age of Teletechnology* (1999), and *The Affective Turn: Theorizing the Social* (2007).

creator or factor of an environment, but is the environment itself. Anthony Giddens's concept of "time-space distanciation" deals with such issues of "environment." Giddens's position begins with the conflation of time and space. Referring to the cultural geographer Torsten Hagerstrand and his colleagues, Giddens charges time-geography with holding on to a principle of social agents as intentional beings with purposes or "projects."[45] Justifiably, he critiques the notion of free agents upheld by the time-geographers and offers his own "regionalization of time-space," defined as movement of life paths through settings of interaction that are spatially demarcated in many different manners.[46] Giddens's criticism extends beyond the "naïve conception," as he calls it, of the human agent, to the environment by introducing a shift from "place" to "locale."[47] "Locale," according to Giddens, refers to the settings of interactions that are crucial to specifying their contextuality and, thus, provide fixity. Therefore, locales self-regionalize; each region constitutes the context of social interaction.

Giddens dismisses time-geography's concept of "place," preferring his own terms, "locale" and "region," because of their implications of presence in time, or temporality.[48] "Presence" entails a level of visibility or detection by an observing party, but not all social phenomena are necessarily detectable through science or the human senses. Not all social facts are knowable in waking life, or reality; Freud's unconscious refuted such claims and presented another space in which social facts lay. Therefore, Giddens's preference for "locale" is in some ways connected to the ontology of Being, referred to by Clough, as a movement away from the Heideggerian ontology of presence.[49] As Giddens notes, the level of presence-availability is not connected to close physical proximity for several reasons, most importantly, as he notes, a change in the one-to-one relationship between the media of communication and the media of transportation. The separation of communication and transportation media he attributes to the development of electronic signaling, which no longer was bound to the mobility of the human body.[50] He concludes, "All social life occurs in, and is constituted by, intersections of presence and absence in the 'fading away' of time and the 'shading off' of space."[51]

However, Giddens's "environment," what he describes as the very "intersections of presence and absence," seems still to be lodged in traditional notions of space. For one thing, absence and presence, like zeroes and ones, is predicated on a dichotomous relationship such as those of modernity—subject/object, human/nonhuman, and so on. Also, presence and absence of an entity requires just that: a solid, whole, unitary entity. Giddens takes on the cultural geographers for having a flawed conceptualization of the subject, or the agent, but I would suggest that Giddens's agent, too, is flawed.

Is something—in this instance, the agent—merely there or not? It seems as though Giddens's "agent" exists in a traditional "structure." However, this cannot be the case with the realities of technomediated infomobility. The Internet's (de)formulation of space affects material space in a way that connects the real with the virtual. Presence as posited by Giddens begs another question: What or who exactly takes up space? Though Giddens offers a useful critique of time-geography, his notion of "environment" and its "agent" leaves much to be desired. Giddens does not go far enough to ask what "matters" in an era of technomedia where space and time are, in themselves and in relation to one another, dynamic. Hence, we must be able to look at space in terms of modulations and dynamic back-and-forth actions where the polar ends of the spectrum are rarely of any significance. Spaces, and environments, are necessarily multiple; the Internet contests our preconceptions of such things.

Folding

In *The Fold*, his reworking of the work of seventeenth-century philosopher Gottfried Leibniz, Deleuze argues over categorization by delving into the possibilities of Leibniz's philosophy to rethink spatial relationships, with particular attention to what can generally be called matter. *The Fold* aims to uproot the idea from traditional definitions with a tendency toward a fixed characterization of what can take up space.

Matter, Deleuze writes, "offers an infinitely porous, spongy or cavernous texture without emptiness, caverns endlessly contained in other caverns."[52] In this rethinking of matter as an assemblage of spaces that gives rise ultimately to other spaces, Deleuze argues against essence or "place" or fixity to matter. There is merely a "pond of matter in which there exist different flows and waves."[53] Matter, Deleuze suggests, can be seen as folds contained within other folds. To make his point clearer, he uses the metaphor of origami.[54] Deleuze explains this image, pointing out that the different folds of the origami are the lines of differentiation that distinguish pieces of matter from one another. Thus, the most important aspect of his notion of folds is that distinctions,

Gilles Deleuze (1925–1995) was a French philosopher whose two volumes *Capitalism and Schizophrenia: Anti-Oedipus* and *A Thousand Plateaus*, co-authored with Félix Guattari, are two of the most widely influential works from the poststructuralist tradition in contemporary social and cultural theory and philosophy and cultural studies. He was a colleague and close friend of Michel Foucault during their time together at the University of Paris VIII-Vincennes/St. Denis, an experimental branch of the well-established city university system in Paris. Their works bore striking conceptual and political affinities and their connection was more demonstrably exemplified by their involvement together in Groupe d'Information sur les Prisons. Deleuze's work is so innovative, in part, due to his rejection of traditional methods of reading classical philosophers. Many of his books are, in fact, radical interpretations of classical, and already well critiqued, philosophers: Kant, Nietzsche, and Bergson, just to name a few. Additionally, his studies of film—*Cinema 1* and *Cinema 2*—are extremely significant, due to his insistence on temporality and movement in cinema. In this book, Deleuze's concepts are in use to contextualize the work of De Landa.

Virtual: The *virtual*, as understood by Deleuze, is always in correspondence with, *not* opposed to, the actual. Though much of the common use of the word refers to things like "virtual reality," Deleuze's formulation of the word is far less negative in connotation. When he describes something as virtual, he is pointing to something that has the potential to actualize, not something that is in opposition to reality, as "virtual reality" tends to connote. Thus, the virtual does not describe the impossible, but rather potential to actualize. It is clear how De Landa applies the virtual, as theorized by Deleuze, to the dimension of matter. In my discussion of matter as that which takes up space, the virtual actualizes using intensive processes that were immanent. Thus, the virtual and actual are both very much real.

such as that between organic and inorganic matter, are not essential. In other words, differences are not natural, mutual exclusions, but are rather results of mechanical differentiations.[55] This mechanical movement, Deleuze is quick to point out, is in no way an exercise of reason. Instead the pleats of matter, the folds, move as if they lack reason.[56] The mechanisms "stretch to infinity because the motivating force is of an already infinite composition."[57] In point of fact, Deleuze is pushing us to think of matter, and thus space, as temporally shifting and contingent.

What Deleuze defines as matter, referring to *folds*, I would like to rethink, in the context of the Internet and technomedia, as spatial *modulations*. Modulations are openings and closings of spaces, but they are never at a fully closed or open state. Therefore, the Internet distinctly challenges the ideal of enclosures, characteristic of the spatial logic that was necessary in industrial capitalism.[58] The Internet, marks not only a new era of capitalism (advanced or late) but also the beginnings of an abandonment of fixity and centeredness of space.

Modulations

The structure of hyperlinks on the Web in many ways demonstrates the break from enclosed space and the movement to folds and modulations. Hyperlinks are designed to connect websites that have some sort of broadly defined affiliation with one another. When I refer to the "structure" of hyperlinks, I am perhaps referring not actually to a structure, but to what Manuel Castells calls a network, a horizontal organizational structure in which the URLs (the websites) are mere nodes (within the network).

In other words, a network comprises multiple nodes with multiple roles. The Web provides a cogent example of Deleuze's folds.

If we propose that hyperlinks are folds, or spatial modulations, it is worth looking into the supposed structure or organization of linked websites to demonstrate such a claim. The "links" section of any given website (or the "blogroll" on blogs) suggests, as stated earlier, a connection between the site and the other sites listed on the "links" page. The connection could be informational (content-specific) or even a social network tie, as many blogs are today on the Internet. Google, the most popular search engine in the world, has its own system of connectivity between websites with its "Similar Pages" function, which is based on content (word-frequency, better known as a keyword search). The appeal, ultimately, for the hyperlink connection lies not solely in the mass of information available from speeding on the (information) superhighway, but in "the stretch to infinity"[59] that hyperlinking offers. Hyperlinking opens up a potentially infinite number of spaces, illustrating precisely Deleuze's point that "a fold always refers to other folds."[60] The website can no longer be seen as a thing in itself, but a folding/unfolding or modulation of information, cultural meanings, signifiers, images, and so on. The explosion of the "text," as Roland Barthes once described, is achieved on the Internet, not to mention the immanent accordance of the two, as we shall see soon enough with the introduction of broadband cable channels such as Al Gore's Current TV, which, much like Mozilla Firefox (web browser), Wikipedia (web encyclopedia), and of course the popular YouTube, is open source or, as in the case of YouTube, user-content driven.

The trend toward an open-source model is inextricably linked to what many call online "piracy" or file-sharing. Starting with the initial hit to file-sharing as a result of the Napster case, there has been a backlash from media conglomerates (as expected) and various state governments globally (in the form of legal action based on Internet copyright laws) against file-sharing of all forms (not strictly limited to audio mp3 files). This backlash can be seen as the constriction of space, or as an attempt by the State and big business to not only protect copyright but, in effect, also hold onto traditional (enclosed or open) formulations of space. The onset of open-source software like that developed by Mozilla, specifically their well-received Thunderbird e-mail client and Firefox web browser, threatens the closed space of web browsers dominated by Microsoft. At the least, it is ironic that Microsoft's Internet Explorer is the main interface that allows users of the Internet to download other web browsers and other e-mail clients like those developed by Mozilla. The market domination of which Microsoft can boast is shaky and always unstable because the Internet is not a traditional space over which one can lay claim; it is, as Deleuze suggests, porous. Its very own instrument of market domination allows for the unfolding of its competitors' products.

The Internet contests the enclosure-oriented ideal of space that remains even in the mainstream vernacular of technomedia and the Internet. This type of language is no longer fit for the present moment. For instance, the "website," as a concept, is misleading and anachronistic. The website, as a space, holds onto its URL (web address) for some semblance of fixity. Yet, its content—the matter of the website as opposed to its state—by design, is supposed to be constantly updated. Its form is, in theory, always changing and "under construction." (Note: Web design, even today after the huge dot.com bust, is a lucrative industry.) Any website is only nominally invested in a fixed identity. News websites (e.g., the sites of CNN, Al Jazeera, or BBC News) never have the exact same information on their home pages from one minute to the next. These websites rely on the fast-paced dissemination of news information. It is contradictory for news websites to have the same content from one minute to the next. In essence, these websites are structured around constant change. Home pages of websites, for better or for worse (though I would say "better" if I had to choose), do not serve as a Cartesian "soul" or essence.

In point of fact, one does not need to check the home page of any website to keep abreast of its changing information content. RSS and XML feeds are some of the recent technologies that allow for live updates. Subscribers can receive feeds to sites that change content frequently (e.g., news websites or music blogs). Mozilla's Firefox web browser integrates RSS feeds from *BBC News Online* through a bookmark called "Latest Headlines." With such a feature, one can receive the content, in this case news information, without having to navigate to the BBC's URL (http://news.bbc.co.uk). Websites always exist beyond their own URLs. This is evidenced by Google's new SMS (short message service, more popularly known as text message) feature. On mobile phones, one only needs to text message any inquiry to 46645 (GOOGL), and Google will respond with information. I am suggesting here that Google or any other website's ability to matter beyond its URL (google.com), demonstrates the spatial modulations of the virtual age.

It is important to acknowledge some of the significant changes occurring as technologies increasingly take hold in the world. Space and time cannot be seen as definite absolute units of measurement because the Internet, in particular, disorganizes and multiplies space. As content and form collapse into one another, the concept of space on the Internet struggles to hold onto fixed identity and centrality, for new technologies rely on a pace that no longer resembles the time signature of modern life. Thus, the Internet and allied technologies unfold and fold our spatial and temporal realities. Multiple tabs on our web browsers, instant messaging, e-mail, stock quotes, pop-up ads: the list goes on and on. The new spaces and temporalities of the Internet generate a semiotic neurosis that dislocates us humans and pushes us to question our own status as creator and master of technology. Information technologies, as they are popularly called, are not merely extractors of resources and knowledge, but are knowledge-production technologies that shape how we view and think of the world and ourselves. In a Borgesian fashion, information has become the environment itself.

Notes

1. Belinda Barnet, "Infomobility and Technics: Some Travel Notes," *CTHEORY* October 27, 2005: 1, at www.ctheory.net/articles.aspx?id=492.
2. GPRS is a data service on mobile phones that allows for text messaging as well as Internet communications such as e-mail and Web browsing.
3. Barnet, "Infomobility," 2.
4. Barnet, "Infomobility," 1.

5. Barnet, "Infomobility," 1.

6. Zygmunt Bauman, "Time and Space Reunited," *Time and Society* 9 (2000): 171. See also chapter 3 of Bauman's already-classic *Liquid Modernity* (Cambridge, UK: Polity Press, 2000).

7. Bauman, "Time and Space," 172.

8. Bauman, "Time and Space," 172.

9. Bauman, "Time and Space," 172.

10. Bauman, "Time and Space," 177.

11. Bauman, "Time and Space," 177.

12. Bauman, "Time and Space," 178.

13. Bauman, "Time and Space," 177.

14. Bauman, "Time and Space," 177.

15. Jacques Derrida, "Signature. Event. Context," in *A Derrida Reader: Between the Blinds*, ed. Jacques Derrida and Peggy Kamuf (New York: Columbia University Press, 1991), 83–84. Emphasis Derrida's.

16. Derrida, "Signature," 85.

17. Derrida, "Signature," 87.

18. Derrida, "Signature," 107.

19. Paul Virilio and James Der Derian, "The State of Emergency," in *The Virilio Reader*, ed. Paul Virilio and James Der Derian (Oxford; Malden, Mass.: Blackwell, 1998), 46.

20. Virilio and Der Derian, "State of Emergency," 46.

21. Virilio and Der Derian, "State of Emergency," 50.

22. Albert Einstein, "On the Idea of Time in Physics," in *Relativity: The Special and General Theory* (New York: Henry Holt, 1920; Bartleby.com, 2000), at www.bartleby.com/173/.

23. Einstein, "Behavior of Clocks and Measuring-Rods on a Rotating Body of Reference," in *Relativity*.

24. Richard Dienst, *Still Life in Real Time: Theory after Television* (Durham, N.C.: Duke University Press, 1994), 167.

25. Dienst, *Still Life*, 159.

26. Dienst, *Still Life*, 160. Emphasis mine.

27. Dienst, *Still Life*, 161. Emphasis Dienst's.

28. Dienst, *Still Life*, 161.
29. Dienst, *Still Life*, 162.
30. Dienst, *Still Life*, 164.
31. Manuel De Landa, "Deleuze, Diagrams, and the Open-Ended Becoming," in *Becomings: Explorations in Time, Memory, and Futures*, ed. E. A. Grosz (Ithaca, N.Y.: Cornell University Press, 1999), 31.
32. De Landa, "Open-Ended Becoming," 29.
33. De Landa, "Open-Ended Becoming," 30.
34. Ian Hacking, *Representing and Intervening: Introductory Topics in the Philosophy of Natural Science* (Cambridge; New York: Cambridge University Press, 1983), 36.
35. Ibid. 32.
36. Immanuel Wallerstein, *The Uncertainties of Knowledge: Politics, History, and Social Change* (Philadelphia: Temple University Press, 2004), 76–77.
37. Wallerstein, *Uncertainties*, 77.
38. De Landa, "Open-Ended Becoming," 35.
39. See Gilles Deleuze, *Difference and Repetition* (New York: Columbia University Press, 1994).
40. Emmanuel Lévinas, *Time and the Other and Additional Essays* (Pittsburgh, Penn.: Duquesne University Press, 1987), 48.
41. xB Browser is an executable file that can be stored on a USB flash drive. When browsing from any computer, one can run xB Browser, which creates a secure, encrypted connection between the computer in use and an exit computer, circumventing any official surveillance. See www.xerobank.com/ for more information.
42. Manuel De Landa, *Intensive Science and Virtual Philosophy* (London; New York: Continuum, 2002), 69.
43. Patricia Ticineto Clough, *Autoaffection: Unconscious Thought in the Age of Teletechnology* (Minneapolis: University of Minnesota Press, 2000), 30.
44. See Howard Rheingold, *The Virtual Community: Homesteading on the Electronic Frontier* (New York: Perseus Books, 1993).

45. Anthony Giddens, *The Constitution of Society: Introduction of the Theory of Structuration* (Berkeley: University of California Press, 1984), 113.
46. Giddens, *Constitution,* 116.
47. Giddens, *Constitution,* 116–118.
48. Giddens, *Constitution,* 118.
49. Clough, *Autoaffection,* 30.
50. Giddens *Constitution,* 123.
51. Giddens, *Constitution,* 132.
52. Gilles Deleuze, *The Fold: Leibniz and the Baroque* (Minneapolis: University of Minnesota Press, 1993), 5.
53. Leibniz as quoted in Deleuze, *Fold,* 5.
54. Deleuze, *Fold,* 6.
55. Deleuze, *Fold,* 6.
56. Deleuze, *Fold,* 10–11.
57. Deleuze, *Fold,* 8.
58. See Michel Foucault, *Discipline and Punish: The Birth of the Prison,* 2nd Vintage Books ed. (New York: Vintage Books, 1995).
59. Deleuze, *Fold,* 8.
60. Deleuze, *Fold,* 8.

4

Ghosts of the Subject

> "For what the unconscious does is to show us the gap through which neurosis recreates a harmony with a real—a real that may well not be determined."
>
> —Jacques Lacan[1]

We were in search of an espresso machine, my friend Jonas and I. He had just spent time in Urbino, Italy, drinking espresso after every meal and wished to relive those moments back in the States. I, on the other hand, had not traveled anywhere, but am an avid coffee drinker—the addiction to caffeine I inherited, I suppose, from my mother (who suffers for it now with hypertension). We needed one for cheap and by convention, we went online, to eBay. Following protocol we typed in "espresso machine" in the keyword search box and found hundreds of espresso machines for auction. We sorted by price and put a bid of $25.00 on a less-than-mediocre brand espresso machine in an auction that was to end in a half-hour.

As the minutes drew nearer to the end of the auction, we entered into a trance of sorts. The dizzying effects of the multiple windows open on the monitor, the web browser on eBay.com being just one of many (instant messenger, e-mail clients, iTunes, etc.), contributed to the collective, if not social, *neurosis* I shared with my friend. "Refresh!" was the incantation as the rush experience peaked as the seconds drew closer to the end. Finally, we did win the espresso machine. However, the satisfaction of winning an auction did not end there. The satisfaction ended when we received the machine in a box within the week and got to sip our first espressos made from the machine we had won on eBay, soaking in the caffeinated joy and experiencing the pleasure of transgressing old rules of space and time through technology.

Social Neurosis: New Modes of Being

In certain surprising ways, technology has also played a hand in distorting and disrupting the very subjects that claim to have created technology in the name of conquering, if not colonizing, nature. In the supposed domination of nature, man's (and I use this gendered pronoun for a reason) modern knowledge (science) represents his utter dominion over all that is seen and unseen. Its narrative, postulating humans as the causal agents of history, though still believed by many, is hauled into crisis and under attack by the very "creations" or objects of the subjects—technology and its subsequent simulacra, the McWorld in which we now live. The subject, or what is left of it, must live within the systems, orderings, and structures of her own creations.

One of these systems is the ever-so-familiar eBay, which has endured the rise and fall of the dot.com boom. Its mission statement reads:

> eBay's mission is to provide a global trading platform where practically anyone can trade practically anything.[2]

Anything? In 2001, Mendi and Keith Obadike, a husband and wife team of interdisciplinary, multimedia artists, offered Keith's "blackness" for sale on eBay,[3] stretching and certainly testing eBay's value-free, market-oriented vision of itself. Though obviously engaged in a political act, maybe even a publicity stunt, Mendi and Keith Obadike did call out eBay's seemingly innocent mission of providing a platform for the trade of goods between individuals. eBay's mission, evocative of the preindustrial marketplace, provides a façade of freedom. Freedom all around, for consumer and capitalist and for buyer and seller. In essence, it bestows a freedom to be your self but also to trade aspects of it. Yet, eBay contains mythic and magical components despite its very "real" ambitions of facilitating the most material and basic capitalist desire of accumulating commodities. In the moments before the end of an auction, the bidder must frantically click to refresh the web browser window repeatedly in order to see if anyone has outbid her. And God forbid anyone has in fact done so, because then there must be a retaliatory action, a violent one of sorts, transfigured into placing a higher bid than the unknown opponent sitting behind another computer potentially anywhere in the wired and increasingly wireless world.

This violence of sorts indicates the degree of affective involvement that occurs between individuals on the Internet on the levels of identity- and subject-formation as well. Following the earlier discussion of Derrida, the Internet cannot be seen as simply a neutral communications media where stable subjects come to communicate with one another. It is, as Charles Lemert and Anthony Elliott describe in *Deadly Worlds*, where the process of individualization and identity-formation occurs:

> But what does arise for consideration, surely, is the issue of how today's imbrication of technology, information and communications alters or transforms not only how we experience, our sense of identity and individualism but also the very process—at once social and emotional—by which young people come to think of themselves *as* individuals.[4]

The concept of collective *neurosis* is indebted to psychoanalyst Jacques Lacan's explanation of cause and causality in Freudian theory in his collection of lectures *The Four Fundamental Concepts of Psycho-Analysis*. I label the frenzy of eBay as *neurosis* for a simple, but still significant and relevant, reason. The unconscious, as interpreted by Lacan, exists at a point where there is something wrong, a disjuncture between a causal force and that which it is supposed to effect.[5] Neurosis, as indicated by the quotation at the beginning of this chapter, "recreates a harmony with *a* real,"[6] which we assume is one of many. Surely, there was something that was breaking up, a fissure, when Jonas and I won the bid for our espresso machine. This particular experience, most likely identified by many as an everyday occurrence, was much more than just shopping. It pointed to a

Jacques Lacan (1901–1981) was a psychoanalyst from France who was responsible for a radical movement in psychoanalysis that broke away from the largely Anglo-dominated scene as exemplified by Erik Erikson. Instead of issues dealing with the conscious self, Lacan was interested in the unconscious, most specifically, the similarities in its structure to language. Lacan's rereading of Freudian psychoanalysis was presented in the form of lectures, *not* the usual academic fare of books and articles. It was highly influential to many who were interested in the linguistic structuralism inspired by Saussure, which had been associated with the likes of Roland Barthes and Claude Levi-Strauss. In fact, his lectures were a focal point for the Parisian intellectual scene; people like Maurice Merleau-Ponty and Jean Wahl frequently attended them and were influenced by them. Today, Lacan's contribution is seen not so much in the field of psychoanalysis, but in social, cultural, and literary theory, which in part is due to the fact that his intellectual heirs have had interests and specialties beyond psychoanalysis.

The Real: It is a popular misconception to read Lacan's *the Real* as connoting reality. In fact, it might suffice to say it is the direct opposite. The Real is but one part in Lacan's schema of Imaginary-Symbolic-Real, which has parallels to the elements of the psychic structure named by Freud as Superego-Ego-Id. For Lacan, the Real cannot be accessed by the subject, which perceives and imagines the visual world in the Imaginary and is then constituted within the field of signifiers into the Symbolic. The Real, as Lacan writes, is that which exists in the "raw"—uncooked by language and visual codes of the world, ultimately only denotable by *x*.

burgeoning new string of relations between consumption, technomedia, and subjectivity.

So what is this neurosis that characterizes the Internet? It is not so much a phenomenon as it is a new universe. It is a new mode of being, as well as a new subjectivity that is not limited to the individual's time spent "online." Whether we like it or not, we are always online. Technomedia undermine the privilege of an offline, "real" existence. A mediated and a "natural" existence are interrelated to the point of confusion.

Guilty as Charged: Shopaholism and Neurosis

In *Civilization and Its Discontents*, Freud describes neurosis as rooted in living in a society steeped in ideals of happiness that are inevitably unreachable. Within this definition of neurosis, he introduces the renowned *pleasure principle*, according to which man moderates his purpose in life—happiness—but settles for the avoidance of unpleasure or suffering.[7] Due to the cultural ideals of civilization, as well as threats of suffering from many directions,[8] the pleasure principle is compromised and the subject works instead to negate suffering, not to fulfill happiness. Thus, the superego, or Culture, seizes the pursuit of pleasure, leaving some individuals to counteract this by engaging in a substitutive activity that induces "a flight of neurotic illness," which Freud interestingly enough calls "media."[9] Although he was alluding to intoxicants, Freud describes the incentives for this "flight" as not only providing immediate pleasure, but "also a greatly desired degree of independence from the external world," by which one is able to "withdraw from the pressure of reality and find refuge in a world of one's own with better conditions of sensibility."[10] Neurosis, he aptly explains, "undermines the modicum of happiness of civilized men."[11]

The Freudian definition of neurosis highlights important issues concerning the psychic orientation of modern subjectivity. It elucidates neurotic practices of everyday life that are utilized in the midst of the frustration of living up to the standards of modern culture. One enters a neurotic state when one cannot withstand the suppression of desire imposed on one in the name of society's cultural ideals. Subjectivity on the Internet in postmodernity operates in a similar fashion, for it is susceptible to both the reality and pleasure principles. Though Freud juxtaposes these two in opposition to one another, I believe subjectivity, irrevocably neurotic, operates through the dual tasks of retreating from the external world (what some theorists have deemed "individuation") and immersing in the technomediated virtual universe. The subject in the virtual universe finds herself at once connected and alienated, leading some scholars to suggest that the reality and pleasure principles have struck a deal.

Just surfing the Web, one has access to information that is unparalleled in human history. This is no surprise. Many view the Internet to have a representational quality to what Freud calls the "external world," that which is "outside of the subject." It is popular convention to do all levels of research on the Web—everything from looking through scientific databases to just simple news searches. To gain knowledge of the world around us, we seek to understand it and inevitably connect to it through technology. This endeavor toward "self-knowledge," Heidegger and Marx remind us, is countered by an alienation from humanity. This tension is furthermore demonstrated by the coupling of fear and ease in parents concerning the Internet-savvy behavior of their children today. The fear of many parents of isolation and antisocial behavior as a result of rising tech-savoir in their children goes hand-in-hand with excitement over the ease of educational engagement with global affairs via the Internet.[12] This fear of children on the Internet stems from a concern about the future of humanity expressed frequently as paternalistic paranoia: "What is happening to our children?" Parents of these postmodern, or what Lemert and Elliott call "technoliterate," children "worry that technoliteracy achieved in childhood is purchased at the cost of a loss of interpersonal maturity due to premature withdrawal from family intimacies."[13] The parents fear that their children will grow up maladjusted to society.

Critics opposed to technomedia, like many parents of technokids in postindustrial countries, are afraid of the effects of technology on "humanity." Neil Postman begrudgingly noted that an era in which there is "uncontrolled growth of technology"[14] brings with it a "new metaphysics" by which the "tie between information and human purpose has been severed."[15] The human subject for Postman is bombarded with information to a point where information appears at random and without purpose or meaning. He fears that information flow has reached a breaking point at which technology is now both means and end to human creativity, stunting "pure" human advancement.[16] It need not be said that Postman fears an eclipse of human reason or a rupture in personal autonomy of the free individual.

The neurosis of the Internet can thus be seen as a locus for a meaningless (as in, without meaning, *not* insignificant) confusion of compulsions, drives, and desires from various sources—internal and external; past and future. This is the import of psychoanalysis, which has always been open to varying sources of subjectivity, in the study of technomedia. Neurosis is, as Lacan writes, a *re-creation* of the harmony with a real. So, is the neurosis of everyday life, entrenched in the mediation of the Internet and other media, an attempt to harmonize, as Lacan suggests? Neurosis is actually not harmonious in the sense of agreement and pleasure, but similar to the harmony envisioned by electronic musicians who create what is called glitch music, utilizing bits and pieces of prerecorded material sampled and tweaked for a new song based on blips and glitches, rearranging rules of tonality, rhythm, and aesthetics.

When those unfamiliar with glitch music encounter it for the first time, many hear it as "mindless noise." Precisely. The derogatory use of "mindless," simply a way of identifying glitch as without reason, regulation, and of course merit, parallels the harmony Lacan speaks of, that of a logic not based on consistency nor agreement, but one that resembles glitch music, unable to be pinned down as calculated behavior or action, but, as Freud suggests, more of a drive. Like glitch, neurosis is the modulating tension of forging identities between the technomediated external world and the individual psyche. It is a mode of existence by which one is able to survive the changing tempo of the virtualized universe.

Visual Consumption and Illusions of a Self

In her groundbreaking 1975 essay "Visual Pleasure and Narrative Cinema," feminist film theorist Laura Mulvey notes that the experience of watching a film, what I call *visual consumption*, directly constitutes the male-identified viewing subject. Mulvey and various other feminist film theorists following her utilized Freudian and Lacanian psychoanalysis to undertake "the analysis of form, [taking] as its point of departure the forming and shaping of the subject in reading and writing, that is, in meaning construction generally."[17] Feminist film criticism, Patricia Ticineto Clough notes, assumes that the film apparatus—the projector, the screen, and indeed the viewer—"[elicits] already existing fantasies of unconscious desire that are condensed both in the individual and [its] social formation." The individual's "knowledge of the world and of self is always constructed in unconscious desire."[18] Though unwittingly, feminist film criticism brought to bear on film Louis Althusser's Lacan-inspired theory of ideological interpellation. Indeed, the process of film-viewing, for feminist film theorists, constitutes the subject of hegemony, who embodies the dominant ideology, which, in this instance, is patri-

Laura Mulvey (1941–) is professor of film and media studies at University of London, Birkbeck. A feminist film theorist, she is famous for critiques of Hollywood cinema using Freud and Lacan. In particular, she argues that the cinematic apparatus places the audience's gaze in the masculine position and thus reproduces the patriarchal, heterosexual social order. Her most famous essay, "Visual Pleasure and Narrative Cinema," can be found in *Visual and Other Pleasures* (1989). She is also a noted filmmaker, collaborating often with Peter Wollen.

archal. The film apparatus presents objects—the stereotypical sexist image of women—to constitute the masculinized gaze of the viewing subject. Thus, the subject's very constitution and sympathetic identification with the male protagonist of the film relies on the voyeuristic and fetishistic objectification of the women characters on the screen.

The experience of watching television, on the other hand, according to Félix Guattari, is a splitting of consciousness. The viewing subject exists in "captivity to the narrative context of the [television] program" with "lateral vigilance to surrounding events."[19] For Guattari, the gaze does not completely take hold of the viewer's consciousness, but rather splits it because of two key differences in technology and media environment. Technologically, the television broadcast does not come from behind the subject like in film. Instead, it comes from behind the screen. This contributes to a difference in media environment. Whereas the cinematic experience forces the viewer to give up her consciousness to what is on screen acting as a replacement to vision itself, television, as Guattari notes, allows for distraction. The form of television programming structures itself around the distraction in the form of the commercial.

Referring to television advertisements, Jean Baudrillard argues that indeed the self was always an effect of the mode of consumption in capitalist societies. "The structures of consumption," he writes, "are simultaneously fluid and enclosed." On one hand, as producer, the individual is incorporated into what Marx called social labor; on the other hand, as consumer, he is solitary and individualized. The individual, or even the mass of individuals, what is generally called "the private sphere," Baudrillard argues, cannot resist "the object of consumption"—the television ad—because it

Jean Baudrillard (1929–2007) was a French theorist whose body of work has ranged from what can be called a "post-Marxist semiotics," as exemplified by his early work *For a Critique of the Political Economy of the Sign*, which was an attempt to reconcile the Marxist theory of value with that of the theory of signs in the vein of Ferdinand Saussure, to more recent works on terrorism and 9/11. It is for this reason, as well as his denouncement of the "reality" of the Persian Gulf War, that many have trouble pinning him down to a "proper" academic field. However, it is clear that he has dealt with *mediation* in almost all of his works. Until his death, he was professor emeritus at the University of Paris. In our discussion, Baudrillard serves as an important theorist of consumption. Two of his terms are of utmost importance to the discussion found in this book:

Hyperreal: Baudrillard describes *hyperreal* as "the meticulous reduplication of the real, preferably through another, reproductive medium." In effect, he is pointing to the nature of media technologies that produce *simulacra* (see below), in which a representation of reality, such as the news and other mainstream media, is superseded by a pure repetition of a "hallucinatory resemblance" to itself. Thus, in contemporary life, what is real or authentic is, in fact, a redoubled version of its own purported authenticity. Ultimately, Baudrillard is pointing to something he calls a "fetishism of the lost object," a mode of knowledge-production in which the commodity of the production process is uncertain and indeed immeasurable for its truth-value.

Simulacra: In Baudrillard's thought, the hyperreal becomes immanent through *simulacra*, a simulation of reality. To give a brief definition of simulacra is highly difficult even for Baudrillard, and thus he presents Disneyland to illustrate his point. Disneyland, though posturing itself as an imaginary enclosed world onto itself, is in fact a miniaturized version of America, Baudrillard says. While attempting to establish the world outside of Disneyland as "real," it acts as a distraction and deterrence from recognizing that reality itself is full of simulations. It not only perverts reality, but hides the absence of one.

is structured by the mode of production, "whose strategy of desire invests *the materiality of our existence with its monotony and distractions.*"[20] Baudrillard suggests that the very existence of individuals is "collectively assigned" in relation to social control, which is heralded, "with increased intensity, as the *reign of freedom.*"[21] Thus, the function of the individual is *ego consumans.*

The Internet acts to subsume and incorporate these two seemingly contradictory technomedia, as well as their social effects. It facilitates a dynamic platform of ever-present and ever-available objects of desire for a postindustrial, consumer society, "splitting" consciousness but also seducing the unconscious, in particular, fantasy. In postmodern societies, if there can still be talk of a subject, quite certainly he or she is a consumer of material goods as well as of technomediated symbolic images, as Baudrillard suggests. Though credit cards and services such as PayPal and Amazon.com's 1-Click Shopping make for easy consuming on the Internet, the availability of information, images, and other data create new spaces from which "machinic transformations of subjectivity"[22] arise, tickling the repressed desires, fantasies lodged in the unconscious, giving rise to a state of neurotic instability.

The Internet hosts many spatial folds on which an economy of identities exists based on the production of both material and visual objects. There are several key points when it comes to the global trade of identities in this mode. First, when speaking of an economy of identities, I am not describing whole persons or entities but, for all intents and purposes, referring to a self that no longer can be thought of as whole and unitary, but a neurotic assemblage of forces, energies, and drives.[23] Furthermore, the consumptive relationship on the Internet does not stand one-to-one with a single consumer consuming a single object.

The Internet's spaces of consumption include the popular and obvious places such as Amazon, eBay, and other retail sites, though they are not limited to them. As these websites strive to replicate commodity exchange in its usual, material form, they also are responsible for a *change* in the concept of consumption and its relation to subjectivity. This type of consumption allows "purchasing" of pieces for an illusion of a complete self. The fragments offered on the Internet are subservient to an impossible, but nevertheless sought after, project of moving toward identity, a predication of a self. Thus, the Internet is the stage, or a field of play,[24] in which consumption is a loosely outlined script. It constitutes all those immersed and subjected in it as consumers caught up in the trade of the various bits and pieces of the fragile, multidimensional sculpture that identity is in postmodernity.

Semiotics of E-shopping

The process of subject-formation and identification involved in consumption on the Internet can be shown through some basic features of a popular shopping website, Amazon (www.amazon.com). Amazon presents itself as oriented toward the personalization of the shopping experience, or what it calls "customer-centrism."[25] Several features on Amazon demonstrate this modification capability. On its main page, Amazon offers many things addressed in the second person, as if to say that Amazon, itself a singular entity, is speaking directly to the consumer (even though it is actually a base for many smaller retailers to sell their products through Amazon's website). The greeting reads:

> Hello. Sign in to get personalized recommendations. New customer? Start here.

"Signing in" on the Internet usually means a registration process. Many websites offer certain services for those users who register with them. These services are, of course, predicated on their ideals of "customer-centrism." An Amazon account includes services such as better shipping deals, deal updates (or junk mail, depending on your level of Internet shopping addiction), and, importantly, recommendations that come in several forms.

When shoppers are ready to purchase any given item on Amazon, their first recommendation comes in a section known as "Better Together," in which Amazon suggests a selection based on genre, author, artist, or any other category that can describe the original item. Just below that, there is "Customers who bought this book [or any item] also bought," which provides links to five or six products that have been bought by other customers along with the item about to be purchased. Within this section, there are also links to other products in other media. For example, if I were in the final moments of purchasing a book on Amazon, there would then be a link to similar items in DVDs and music. Correspondingly, there is another section below entitled "Customers who viewed this book [or any item] also viewed." And below that are two sections that have to do with personal taste: "Listmania!" and "So You'd Like to . . ." Each allows the consumer to view (and create) lists or guides expressing opinions of Amazon users—in the case of books, about what are "must-reads." There are guides, lists, and recommendations galore on Amazon, which is hoping to act as not only an "e-commerce platform,"[26] but also a field of play for a self-formation/affiliation, in Donna Haraway's sense, in commodity purchases.[27]

Looking at the propensity for lists and compartmentalization, Amazon seems to be quite similar to Baudrillard's description of the mall or the shopping center, which he calls "the total organization of everyday life."[28] To extend his argument, I would add that the enumeration logic that retailers on the Internet have undertaken wholly organizes and homogenizes the self on Amazon, which Baudrillard no doubt would characterize as a simulacrum of a mall—a doubly hyperreal version of an already hyperrealized experience. Within the mall, Baudrillard notes, the individual is placed in a world of objects, an artificial environment that is "completely climatized, furnished, and culturalized."[29] When one looks at Amazon's website, the high level of organization is striking. At the top are tabs, on the sides are the several product categories for easy browsing, and smack dab in the center is a section called "Recommended for You," in which several products are selected based on one's past browsing and shopping habits. Though rationalizing in its logic, Amazon entangles several different spheres of life, including work, leisure, nature, and culture—"separate and more or less irreducible activities that produced anxiety, and complexity in our real life, and in our 'anarchic and archaic' cities, have finally become mixed, messaged, climate controlled and domesticated into the simple activity of perpetual shopping."[30] The subject's constitution as individual is no longer grounded in the drawing out of the desire to shop, but of the rationalizing process of identification itself.

Thus, Baudrillard argues that consumptive need is conditioned by one's environment. According to both neoclassical and liberal economists, the economy brings out the natural economic principle in the individual, allowing for harmony with nature. Their theories differ with respect to the individual's relation to the economy. For liberals, the economy imposes on the individual's true objectives, while, for the neoclassicists, the individual exercises rational choice when taking part in the economy in the act of consuming.[31] The latter image of the consumer is of course *homo economicus*, a view of the human as a rational actor who functions based on utility and self-interest. However, consumption practices on the Internet show us otherwise. Consumption in this case is an irrational act, meaning that it does not come from within the deep and authentic interiority of the individual, nor his rationally economic consciousness. Following Baudrillard's line of argument, Amazon has just as much, if not more, to do with the act of consumption than with the individual's own tastes or other conscious thoughts. The online retailer's "1-Click Shopping" feature, in which the customer, having already signed in and thus accessed her already chosen preferred method of payment, is able to buy a product with a literal single click of her mouse, paves the way for impulse-buying. Indeed, Baudrillard's point about the mall or shopping center has imploded, to use his own metaphor; today the post-modern consumer shops away on Amazon without experiencing the climate-control of the building but simply the buttons of the mouse and keyboard.

Subjectivity as Enumerative Individualism

Judging from Amazon, the subject on the Internet is embedded well within a list-oriented and compartmentalized consumer context; it seems that there must necessarily be a look into what exactly lists do for subjectivity. Those who fear the effects of technology, like Baudrillard and George Ritzer, take the position that the "listmania" of the Internet reflects the ultra-rationalization propensity of late capitalism extending to subjectivities and identities. To a certain degree, they are not wrong in their indictment of the increase of consumer activity that comes along with greater degrees of technomedia saturation.

The idea of ultra-rationalization, along the lines of Baudrillard, stems from the argument that even consumer need is conditioned by society. This is found most prominently in the work of the Frankfurt School of Critical Theory. Herbert Marcuse, affiliated with this school, suggests that the one-dimensionality of character in late-capitalist societies has created false needs distributed through mass media and advertising in particular. What upset Marcuse was the decline in revolutionary potential of those living under Western capitalism and Soviet-style socialism, which he saw as a direct consequence of the promotion of consumerism in media. Yet, it is not merely that the ultra-rationalization argument characterizes Internet consumption in particular, but

Herbert Marcuse (1898–1979) is a Marxist sociologist and philosopher who studied with Martin Heidegger at University of Freiburg and joined the Frankfurt School of Critical Theory in 1933. **The Frankfurt School** is the moniker for the Institute for Social Research at the University of Frankfurt am Main, headed by Max Horkheimer, with which several exceptional social thinkers the likes of Erich Fromm, Theodor Adorno, Jurgen Habermas, and Walter Benjamin were associated. Marcuse, having settled in the United States after fleeing Nazi Germany, taught at Columbia University, Harvard University, Brandeis University, and the University of California, San Diego. In books such as *Eros and Civilization* (1955) and *One-Dimensional Man* (1964), he engages in a critique of capitalism addressing issues of science, technology, and desire, taking on, among others, Marx and Freud. More than any other member of the Frankfurt School, Marcuse is seen as most connected to the New Left movements of the 1960s due to his frequent appearances at student protests.

all activity. In fact, a claim supporting such an argument would be that Amazon has created a singular path, one in which the consumer, faced with only so many options, is presented with a mere apparition of choice, producing desires that are only fulfilled by the subject's consumption of objects. Thus, Internet shopping, like any other shopping, becomes an activity fueled by capital and only satisfied by a "thrownness," to use Heidegger again, into a subject-consumer position. The path of happiness becomes that of the mainstream (buying the products on the lists on Amazon), to essentially "fit in." Opinion generators are plentiful, the most noted being magazines (e.g., *Cosmopolitan*, *Glamour*, *Details*, *Esquire*, etc.). And they, though not trendsetters, are responsible for the proliferation of opinions into mass culture. This is a point made frequently by many cultural critics who argue that what is felt by the subject as agency in consumer choice is predetermined, to quite a degree, by mass culture. In today's world, the production of mainstream cultural (and of course political) values is streamlined into an increasingly traceable singular entity. Within the reasoning of these rationalization critics, the predictions of Adorno and Horkheimer could be seen as accurate: The culture industry's increased conglomeration points to the triumph of authority; meanwhile the masses fail to resist the temptations of pop culture, consumerism, decadence, and the desire for bourgeois ideals.

So then, what is the satisfaction or fulfillment derived from consumerism? If indeed consumerism entraps us and routinizes all human activity in the name of the capitalist ideology, why do we find such enjoyment in it? Is it linked to pleasure as espoused by psychoanalytic theory or is it simply the accumulation and production of cultural commodities? Are pleasure and material accumulation even separable today?

Jean Baudrillard's analysis of the shopping mall is crucial here to understand the blurred lines between material consumption, pleasure, and identity. He argues that the mall or shopping center compartmentalizes all spheres of everyday life. The mall influenced what he deems the *disintegration* of the commodity into sign. Hence, consumption changes from buying goods for material needs into what he deems a general hysteria, in which consumption is "not a need for a particular object as much as it is a 'need' for difference (*the desire for social meaning*), only then will we [consumers] understand that satisfaction can never be *fulfilled*."[32] Consumption is not material accumulation simply, but also the action predicated from the desire for meaning. Hence, consumption, in this context, cannot be viewed as simply a system of commodity exchange but also a flow of symbols and signs.

The link between consumption and pleasure in a technomediated milieu is complex to a degree that a clear politics is difficult to derive. Even cultural critic Naomi Klein, author of *No Logo*, admits to "the deep limitations of consumer activism," citing the fact that "[o]ur economic system makes it almost impossible to consume 'ethically,' since everything that is produced within it is produced through the exploitation of human labour and of the environment."[33] This position is a clear retreat from that which was stated *No Logo*, published in 2000 after the antiglobalization demonstrations in Seattle, in which she condemns crass consumerism, claiming that brand names make individuals into advertisements for corporations. Klein's retreat suggests that the subject's very constitution throws her (the subject) into an inevitable project of self-fulfillment, human prosperity, and modernization, albeit transvalued and assigned to a notion of bettering herself by purchasing goods of a particular brand. (The rapper Nas in "Nas Is Like" raps "I went from Seiko to Rolex" to indicate his success.) Elliott and Lemert, in a different way, recognize a similar link between individual identity and self-improvement. "Identities," they write, ". . . have taken the modern individualist impulse for self-constitution and self-improvement and raised it to the second power, thus giving a radical postmodern slant on choosing, changing and transforming as ends in themselves."[34]

For the so-called subject today, the body, mind, and everything in between becomes a point for improvement, a far cry from the stable individual of modernity. The increase in demand for personal trainers, the rise in new diets (Zone, Atkins, and South Beach) and the emergence of a new class of professionals called "life organizers" are just a few indicators of this drive for enhancement and change. Membership in this club seems to be a double-edged sword, considering that being "interpellated" according to the rules of the dominant ideology leads ultimately and almost deterministically to not being "good enough," what Slavoj Zizek deems the *symptom* of contemporary social life. It seems that one's entrance into the global game of identity-formation necessarily means having an identity characterized by multiplicity and complexity, one "with a wondrous capacity for continual change and instant formation."[35]

The criticism of consumerism as exemplified by the Klein of *No Logo* shares similarities to the model set forth by the Frankfurt school and Marcuse. Media, in its function as the booster for consumerist ideology, interpellates subjects down to the level of desire and fantasy. For Klein, mass media was just a forum for brand-name advertising, another node in the network of the culture industries of multinational corporations. The Internet, however, contests such a claim due its spatial logic. As I described earlier, the Internet is a space of spaces, dictated not by hard boundaries, but characterized by links. One space is a gateway to another space and so forth. Thus, it is becoming increasingly difficult to define "mainstream" in the age of the Internet. For one, the phenomenon of music and fashion blogs makes a strong case for the Internet as a contested cultural terrain. Blogs such as theFader.com and ViceLand.com, not to mention the slew of personal blogs generated from Blogspot.com and other blogging host sites, suggest that the Internet offers alternatives to the culture industry.

Furthermore, as Lemert and Elliott suggest, individual identity is *not* singular, most especially because of the technological milieu of the world today. Klein and the other theorists of rationalization miss the point made by Postman, which, though quite one-sided and overly moralistic, has empirical import. The human subject *is* bombarded with information, images, sound, affects, and so on, none of which are organized within the psyche. Though Amazon may resemble the hyperorganized shopping mall, as Baudrillard suggests, the various symbols and images found there do not organize themselves within individual psyches as a stable subject. This analysis not only fails to explore the possibility for multiple identifications to occur, but also overestimates the equivocality of subject-formation in technomedia.

Interpassive Subjectility: The Case of Lurking

Derrida's *subjectile* provides a conceptual starting point for looking at the so-called subject as lost or absent and ultimately resembling a ghost, not singularly an affective (or immaterial) producer or a ready-made consumer. The subject of the Internet, due to its lost and exilic state, acts as tension against the serial identity or enumerative individualism that is facilitated by the hyper–shopping mall character of the Internet. The subjectile allows a look into subjectivity on the Internet as anything but an ordered and linear organizational whole, but instead an assemblage, as Deleuze and Guattari suggested, or series of traces, as Derrida would have said.

The *subjectile*, for Derrida, is "the thing that can take the place of the subject or of the object—being neither one nor the other."[36] It is porous, able to be traversed so as to act like, he describes, "a subject without a subject,"[37] betraying us as it "comes across as being someone instead or something else."[38] It is an apparition of sorts. Its porosity is linked to another facet of subjectility, namely movement or motion. Derrida describes the traversal of the subjectile as spurting (*jetée*):

> the movement that, without ever being itself at the origin, is modalized and disperses itself in the trajectories of the objective, the subjective, the projectile, introjection, objection, dejection, abjection and so on.[39]

On discussion boards, social networking sites, and blogs, there are those called "lurkers." Lurking, in Netlingo, connotes "a visitor to a newsgroup or chat room who only reads other people's posts, never posting his or her own, thus remaining anonymous."[40] Ellen Zweig describes the experience of lurking as reading as opposed to writing.[41] The act of writing, equated with authorship and subjectivity, or in this case submitting a post to a message board, establishes Being for a subject on the Internet. Yet, message boards do not require the subjectivization of writing. Zweig writes:

> All things that seem important, that define us, elude our definition of this creature [the lurker]. This one has no identifying marks, no definite characteristics. We know there's someone out there; we might even know where, how many. We know this one must have gender and race, ethnic identity, tastes and opinions, knowledge and hopes and fears: all of the things that define us. We might compute a statistic probability and create a typical lurker. But as an individual, this one is elusive. We resort to the deictic. There is this one, the unmarked, only with the potential to be marked. The chalk mark on the ground where the body once was.[42]

The lurker exemplifies the subjectile *par excellence* on the Internet. While keeping in furtive perpetual motion, the lurker is not forced to establish itself as a subject. It exists purely as a reader, more passive than active with regard to the discussion. Its non-presence is latent, an "ominous threat" of "pure potential."[43] Achieving subjecthood, in the case of lurking, would be posting on a discussion board or newsgroup, hypothetically breaking out of anonymity by giving oneself a voice, a presence or, in this instance, text.

However, even in the case of self-identification, thus claiming a subject-position (by submitting a post on a discussion board), tricks and deceptions are in play that destabilize the aforementioned subjectivization. The lurker, as is the Derridean subjectile, is a trickster. (The subjectile, Derrida notes, always betrays its observer, posing as the subject.) Internet discussion boards, for example, rarely display the names of their registered users. Most of the time, boards require user names or handles as opposed to birth names. The use of "screen names" undermines the ability to represent a singular "true self." In fact, it is common for an individual to have multiple e-mail addresses as it is for him or her to have multiple aliases on different spaces on the Web. Blogs, for example, do not require an e-mail address or a user name at all to post commentary. When a comment is posted without a name or e-mail address, the default byline on many blogs reads: "posted by anonymous." Lurking, thus, queers the notion of participation and presence. Participation on the Internet does not include an act of "outing," or exposing, anyone. Without having to take responsibility for what he or she says on discussion boards or blogs, the lurker or subjectile undermines the traditional rules of membership in a community because he or she cannot be held responsible for what he or she posts. Lurking makes governance of communities on the Internet quite difficult. The anonymous post is that chalk mark of which Zweig speaks; it is a trace or a memory barely indicating the subjectile's invisible presence. The lurking subjectile of the Internet breaks free from fixed "presence" and participates even as it does not undergo the old processes of entering the social.

Slavoj Zizek makes a similar point when he substitutes "interactivity" with *interpassivity* for the status of the subject in what he calls "new electronic media."[44] According to him, interactivity overstates the "dialogic relationship" between user and object. Obviously unconvinced of the democratic potentials of cyberspace, Zizek argues that in technomedia, the subject is *interpassive*, deferring "*jouissance* to the Other, who passively endures it (laughs, suffers, enjoys) on my [the subject's] behalf."[45] He demonstrates the interpassivity by pointing to Lacan's notion of the *barred subject*, described as "the pure void without substantial content." The barred subject becomes caught in the modulation of folds, riding

Slavoj Zizek (1949–) is a Slovenian philosopher, cultural critic, and psychoanalyst. The sheer breadth of his intellectual output in the past decade includes several widely discussed articles and books, including *The Sublime Object of Ideology* (1989), *The Ticklish Subject* (1999), and most recently *The Parallax View* (2006). Zizek is peculiar in that his writing wavers from high theory to pop-cultural criticism. He meditates on Lacan's relation to Hegel in one article and critiques American television shows in another. This rather schizophrenic nature of his intellectual pursuits explains his great popularity among young academics from a variety of disciplines. His theoretical framework in all of his scholarship is self-avowedly Lacanian. He is senior researcher at the Institute of Social Sciences at the University of Ljubljana, Slovenia, and international director of Birkbeck Institute of Humanities at Birkbeck, University of London.

the liquidity of the Web, struggling to stay afloat, seeking to be fulfilled by a "home" where one is her true "self." The interpassive or barred subject is analogous to Derrida's description of the subjectile. Both are moving toward the conceptualization of subjectivity as not stable but decidedly neurotic and thus empty or without substance. It resembles, as Ellen Zweig writes, a chalk mark on the ground. This barred characteristic is no less evident in the recent strategies of surveillance that stem from the reissue of the Patriot Act in the United States. The U.S. government's new form of surveillance rests upon tracking Internet searches, not phone-tapping or human surveillance grounded in physical locations of bodies. When the Justice department demanded that Google and other popular search sites (such as

Yahoo!) hand over search data, Google, unlike Yahoo!, refused to comply.[16] In effect, the department, in demanding Google hand over search data, has acknowledged that the Foucauldian Panopticon mode of surveillance is no longer applicable when attempting to police bodies in technomedia. The Department of Justice must start with the words that are searched, the very things that are "between above and below, visible and invisible, before and behind," not the individuals themselves per se.

The unusual space of the Internet has facilitated not only the withering of space, but also the disintegration of the interactive user heralded early on by tech and communications firms. Disappointingly (for those interested in holding onto the idea of a true self), there is nothing of the sort except the social neurosis characteristic of the present moment. The neurosis throws what was once the subject, its ghost—the interpassive subjectile—in play into the cacophony of postmodern life, where old rules of cause and effect are transgressed, the aura of technology stripped, and identities are traded in a context of desire/sign/need production. Our present teletechnological moment ensures a drastic transformation in subjectivity and identity. In the aftermath of the dissolution of the subject into liquidity, there are only the ripples of its once presence, only traces of its frantic motion.

Notes

1. Jacques Lacan, *The Four Fundamental Concepts of Psycho-Analysis* (New York; London: W. W. Norton, 1998), 22.
2. eBay, "The Company," at pages.ebay.com/aboutebay/thecompany/companyoverview.html, accessed on March 10, 2006.
3. See the Obadikes' website, www.blacknetart.com/.

4. Charles Lemert and Anthony Elliott, *Deadly Worlds: The Emotional Costs of Globalization* (Lanham, Md.: Rowman & Littlefield, 2006), 26.

5. Lacan, *Psycho-Analysis,* 22.

6. Lacan, *Psycho-Analysis,* 22. Emphasis mine.

7. Sigmund Freud, *Civilization and Its Discontents* (New York: W. W. Norton, 1961), 26.

8. Freud states that there are specifically three directions from which suffering threatens man: (1) one's own body (physical finitude), (2) the external world, and (3) social relations.

9. Freud, *Civilization*, 28.

10. Freud, *Civilization*, 28.

11. Freud, *Civilization*, 39.

12. "Internet Helping with Homework," *BBC News Online*, September 3, 2001, at news.bbc.co.uk/2/hi/uk_news/education/1523404.stm.

13. Lemert and Elliott, *Deadly Worlds*, 22.

14. Neil Postman, *Technopoly: The Surrender of Culture to Technology*, 1st ed. (New York: Knopf, 1992), xii.

15. Postman, *Technopoly*, 70.

16. Postman, *Technopoly*, 70.

17. Patricia Ticineto Clough, *Feminist Thought: Desire, Power, and Academic Discourse* (Oxford; Cambridge, Mass.: Blackwell, 1994), 76.

18. Clough, *Feminist Thought*, 77.

19. Félix Guattari, "Subjectivities: For Better and for Worse," in *The Guattari Reader*, Félix Guattari and Gary Genosko, Blackwell Readers (Oxford; Cambridge, Mass.: Blackwell, 1996), 199.

20. Jean Baudrillard, "The Consumer Society," in *Jean Baudrillard: Selected Writings*, Jean Baudrillard and Mark Poster (Stanford, Calif.: Stanford University Press, 1988), 54. Emphasis mine.

21. Baudrillard, "Consumer Society," 53.

22. Guattari, "Subjectivities," 193.

23. For an excellent overview of theories of the self, see Anthony Elliott, *Concepts of the Self* (Oxford: Polity Press, 2001).

24. Baudrillard, "Consumer Society," 44.
25. Amazon.com, "Media Kit: Overview," at phx.corporate-ir.net/phoenix.zhtml?c=176060&p=irol-mediaKit, accessed on April 2, 2006.
26. Amazon.com, "Media Kit."
27. Donna Haraway, "A Manifesto for Cyborgs: Science, Technology and Socialist Feminism in the 1980s," *Socialist Review* no. 80 (1985): 65–108.
28. Baudrillard, "Consumer Society," 34.
29. Baudrillard, "Consumer Society," 33.
30. Baudrillard, "Consumer Society," 34.
31. Baudrillard, "Consumer Society," 43.
32. Baudrillard, "Consumer Society," 44–45. Emphasis Baudrillard's.
33. Naomi Klein, "Globalization FAQ," www.nologo.org/, accessed February 18, 2006.
34. Lemert and Elliott, *Deadly Worlds*, 86.
35. Lemert and Elliott, *Deadly Worlds*, 86.
36. Jacques Derrida, "Maddening the Subjectile," in "Boundaries: Writing and Drawing," *Yale French Studies* no. 84 (1994): 154.
37. Derrida, "Subjectile,"160.
38. Derrida, "Subjectile,"157.
39. Derrida, "Subjectile,"167.
40. NetLingo: The Internet Dictionary, at netlingo.com/lookup.cfm?term=lurker.
41. Ellen Zweig, "The Lurker: Outline for a Murder Mystery," in *Being Online: Net Subjectivity* ed. Alan Sondheim (New York: Lusitania, 1996), 26.
42. Zweig, "Lurker," 27.
43. Zweig, "Lurker," 27.
44. Slavoj Zizek, "The Interpassive Subject," *Traverses* (1998), at www.lacan.com/zizek-pompidou.htm.
45. Zizek, "Interpassive Subject."
46. "Google Defies U.S. over Search Data," *BBC News Online*, January 20, 2006, at news.bbc.co.uk/2/hi/technology/4630694.stm.

5

Culture, Information, and Politics

> "There is some agreement that the older modernism [high modernist artistic movements] functioned against its society in ways which are variously described as critical, negative, contestatory, subversive, oppositional and the like. Can anything of the sort be affirmed about postmodernism and its social moment?"
>
> —Fredric Jameson[1]

The Politics of Postmodern Culture

Changes in media technologies have always had cultural and political implications. Print culture in the sixteenth century, as noted earlier, provided the groundwork for modernity's claim of a bourgeois public sphere bound together by rationality.[2] Bourgeois, in this instance and in most others, has dual meaning. On one hand, it is clearly a cultural marker, indicating the tendency for its members to share similar values, norms, and customs. On the other hand, as Marx saw it, the bourgeois was the new (for his time) ruling class.

In postmodernity, the shape and speed of culture, its production, and its dissemination have changed tremendously due to the rapid pace of technological innovation and ubiquitous presence of media. Culture, in the most general sense, as it mediates realities, is a knowledge- and value-system that defines the possibilities of the world and even the universe. Accordingly, culture today, as it is inevitably technomediated, is less "real," but more powerful and influential "in the shaping of one's sense of herself and her place in the world."[3] In its detachment from a supposed empirical reality, culture reveals the vulnerabilities, insecurities, and taboos of modern life, now at a pace unparalleled.

The Internet has become the primary environment for cultural articulations of all kinds, playing host to not one dominant mode of culture or cultural production, but a wide variation. Yet, Fredric Jameson and other post-Marxist

Fredric Jameson (1934–) is William A. Lane Professor of Comparative Literature and Romance Studies at Duke University. He is one of today's foremost Marxist literary critics and cultural theorists. He has written numerous books on topics ranging from film criticism to dialectical method. Most of his works can be characterized as injecting a political perspective into interpretations of cultural products. His *Postmodernism or, The Cultural Logic of Late Capitalism* (1991) is already a classic text in understanding ideas of postmodernism and postmodernity.

scholars (e.g., Italian scholars of the Autonomia tradition, most notably Antonio Negri) suggest that, though an authoritative culture has been resisted, there is a common thread weaving together the methods and products of postmodern cultural production nonetheless.[4] According to Jameson, postmodernism is the cultural logic of what had previously been thought of as a fundamentally economic order. He notes that the differentiation of economic and cultural orders is a misleading analysis of capitalism, and this new circumstance changes the face of culture and its function in late capitalism. He argues that, as a cultural dominant, postmodernism blurs the difference between "aesthetic production" and "commodity production." Jameson describes this process as

> the frantic economic urgency of producing fresh waves of ever more novel-seeming goods (from clothing to airplanes), at ever greater rates of turnover, [that] now assigns an increasingly essential structural function and position to aesthetic innovation and experimentation.[5]

Jameson's argument here is a reapplication of Marx's famous line about commodity fetishism: Everything solid melts into air. Commodity production subsumes art or culture, and cultural production exists inescapably within a capitalist mode of production. Yet the issue of the changes in the dynamics of social relations as dictated by culture and its mode of production remains. Culture's interimplication with capitalism and technomedia in the mass social transformation known as globalization brings the relation of culture and production up for discussion. Jameson's argument concerning cultural production must be reconsidered in the case of the Internet. Specifically, I would like to argue that the Internet alters the dynamics of cultural production as well as *reproduction* within what Tiziana Terranova calls a new cultural and informational milieu.

Did Someone Say Capital?

Terranova raises a question that haunts Jameson's periodization of postmodernism and its positing of the complete subsumption of culture to capital: "Is it possible to wage a struggle around culture if all culture has become an industry of signification—drowning in meaning in a sea of semi-random noise?" The tautological formulation that equates culture to industry, Terranova suggests, requires meaning to have simply disappeared, whereas she believes it has "multiplied and proliferated."[6] In light of her criticism, Jameson's argument seems to be quite limited and unaware of the changed (and undoubtedly continually changing) dynamics of information in technomedia. Information, she argues, is not merely the content of communication, nor is it disconnected from materiality (or substance).[7] The categorical distinctions between information and substance are ill-suited for a context in which space and time are reconstituted by technomedia. Terranova sharply criticizes the Marxist (historical materialist) understanding of communications technologies and new media that still holds onto these distinctions.[8] She notes that the presumption that society is solely driven by class struggle and antagonism between capital and labor comes into question when one takes seriously the effects of globalization and of the technological revolution of the late twentieth century.

For Terranova, the nineteenth-century separation of materiality and immateriality, the uncritical notion of infor-

Tiziana Terranova (1967–) teaches at Università degli Studi di Napoli "L'Orientale." She researches in the areas informational dynamics, communications theory, affect, and artificial life. She is author of *Corpi Nella Rete* (1996) and *Network Culture* (2004). She frequently collaborates with Luciana Parisi; one of their informative and exciting texts is "Heat Death" (*CTHEORY*, 2000).

mation as simply "meaning," and the commodification of culture into the forces of capitalist production are awkward applications of Marx's labor theory of value. Such a separation, maintained by Jameson, falls into a position of viewing history as linear and restricts a proper analysis of contemporary capitalism. Furthermore, she adds that the changes in the informational dynamics of globalization are overlooked. She notes that transformations in the mode of production, new forms of power/knowledge, and new informational dynamics link together in a "crowded and uneven communication milieu where global, national, regional and local television networks resonate and interfere with each other and other media such as radio, books, telephony, the press and the Internet."[9]

Production and Reproduction in the Informational Milieu

In this communicative milieu, cultural and political processes seem to have altered significantly in their informational tactics, the very means of communication. The shift in cultural processes, noted by Terranova, is a shift in emphasis from production to reproduction. On this point, she and Jameson are very much in agreement, in that they both understand culture to be a reproductive force, but for Jameson the object of reproduction is the capitalist logic. However, Terranova maintains that there is a conflict within the relations and forces of production. Though Jameson is seen as a decidedly Marxist critic, it seems that Terranova, in this instance, is more faithful. In *A Contribution to the Critique of Political Economy*, Marx writes:

> The mode of production of material life conditions the general process of social, political and intellectual life. It is not the consciousness of men that determines their existence, but their social existence that determines their consciousness. At a certain stage of development, *the material productive forces of society come into conflict with the existing relations of production or—this merely expresses the same thing in legal terms—with the property relations within the framework of which they have operated hitherto.*[10]

Here, we find the famous lines where Marx argues that social existence itself is constituted by the relationship between the forces and relations of production in a capitalist system. As he notes, the forces of production are material, while the relations are, in effect, immaterial. But the lines that are of interest here are those that come after the ones making existential claims. Indeed, Marx realizes the potential for conflict between "the forces of society" and "relations of production." The example he utilizes is property relations. He is referring to the historical shift of property relations from the feudal commons to enclosures, a movement that privatized land into the contemporary notion of private property. In the age of information, the landscape has changed quite a bit, presenting a similar situation to Marx's.

The mode of production in late capitalism has shifted to what Mark Poster calls the mode of information. With this label, Poster alludes to the bestowment of "current culture's fetishistic importance" to "information."[11] Therefore, cultural production of information has fashioned "an unavoidable context of discursive totalization."[12] His analysis is similar to that of Terranova in that they both emphasize a shift in the method of communication, while questioning the concept of *information*. Though disagreeing almost completely on what information means, Poster and Terranova agree that the configuration of communication creates new formations of language and consequently alters the network of social relations.[13] If there is a new mode, then there must also be a new relationship that exists between the forces and relations of production.

Most importantly, Marx's emphasis on production must be reassessed. In the mode of information, the focus when looking at capitalism is not on production, but rather on *reproduction*. The techniques of reproduction have changed too. The cost of reproduction in the digital age has become increasingly more difficult to calculate. It is clear that modern technology sought to reduce the costs of reproduction. The factory system, automation, and Taylorism are all residues of such methods. Moreover, in the new information economy, many scholars, notably Maurizio Lazzarato, have claimed that labor has changed to the production of immaterial commodities by immaterial labor. In a digitalized workplace, the labor-time required for reproduction drastically decreases. Digital reproduction of a commodity does not require much time, if any; hence, the argument

Maurizio Lazzarato is an independent social theorist and philosopher based in Paris. He is known in the English-speaking world for his essay "Immaterial Labor," which appeared in an edited collection called *Radical Thought in Italy* (1996). He writes on various themes related to work, labor, and technologies and also the work of early French sociologist Gabriel Tarde. Lazzarato is affiliated with the monthly review *Multitudes* and also with the earlier *Futur Antérieur*, which was started by Antonio Negri. His notion of "immaterial labor" is explored in detail in the following section.

maintains that labor-time can no longer be used to valuate commodities. Marx's labor theory of value privileges *human* labor-time in commodity valuation, though Marx did indeed understand the importance of technological innovation. With digital technology, it has become clear that nearly no labor-time is needed to produce or reproduce many commodities. As "labor-time" diminishes in importance, what happens to labor, commodity value, production, and reproduction?

Changed Labor

Digitalization, according to Lazzarato, has affected labor as well as commodity production. He argues that this change can be seen in immaterial labor, "defined as the labor that produces the informational and cultural content of the commodity."[14] Work, even the work of bootlegging, has changed substantially; labor is increasingly intellectual and affective.[15] Even so-called blue collar jobs have taken a turn toward what he calls "knowledgeable subjectivities." He notes that in terms of "workers' labor processes in big companies in the industrial and tertiary sectors . . . [the] skills involved in direct labor are increasingly skills involving cybernetics and computer control."[16] Lazzarato states that a new "mass intellectuality" has come into being, resulting in a "new nature of productive activity." Work today, he concludes, involves a "series of activities that are not normally recognized as work."

And due to this change in labor, the worker, now an interface for communication, is more susceptible to the *organization* and *command* of capital.[17] As Foucault had already suggested about the body as a technology of discipline for the social order, Lazzarato follows by making a similar argument about the worker. Offered autonomy or "freedom" in the work place in a nicely packaged subjectivity based on "decision-making"; the worker simply acts a switchboard for the growing complexities of capitalist profit-making. "The new slogan of Western societies," Lazzarato explains, "is that we should all 'become subjects.'"[18] This new subjectivity described by Lazzarato is a Foucauldian nightmare grounded on utilizing the subjective processes of the new worker as participative management, with its main source of power being the investment of the worker's constitution as a subject, with decision-making abilities and perhaps a bigger slice of the pie. Starbucks, as do many other giant corporations, offers shares to its employees and even calls them, rather patronizingly, "partners."

However, Lazzarato continues, "capital wants a situation where command resides within the subject him or herself, and within a communicative process."[19] Lazzarato argues that this new generation of immaterial workers is now hoodwinked into an ideological environment that "constructs the consumer/communicator" relationship as "active."[20] In other words, labor's immaterial character allows for capital to "satisfy a demand by the consumer and at the same time establish that demand." He considers immaterial labor to be "capitalist valorization," an invasion of capitalist production into our lives and the breakdown of "all oppositions among economy, power and knowledge."[21]

In explaining the changing nature of work in what Scott Lash and John Urry call "disorganized" capitalism, Lazzarato laments the confusion of work and play. Using language similar to Habermas, Marcuse, Benjamin, and Lukács, he argues that capitalism has effectively infected our lives by deceiving us into a work-oriented subjectivity that ultimately does the dirty work of creating surplus value for capital, though with supposed benefits. While it seems that work has become easier and less physically demanding, he argues that the ease experienced by workers who are able to use a keyboard and computer monitor or labor affectively in the service industry, for example, is actually just a method by capital to increase work efficiency in the mode of production. Lazzarato seems to prefer an older, more modern capitalism where one could see more clearly the difference between the good guys and bad guys in constructing one's politics. But I wonder whether an older capitalism facilitates a better situation for a cultural politics. One of Lazzarato's main fears (understandably so) is that capitalism has come to be so dynamic and ubiquitous that the forces of production have subsumed all aspects of human life itself. Ultimately, he suggests that there is no individual choice; even the act of consumption is a socially conditioned process inscribed within the logic of a new type of capitalism.

The Politics of Downloading

In various stages of capitalism, the juridical-legal system of the nation-state enforced relations of space, predicated on ideals of private property, to ensure the development of a commodity—land, for example—to generate capital. Thus, a single commodity could only have one owner, but several users, taking the form of renters. Property laws are spatial, and indeed social, relations that are sanctioned by the state and undoubtedly lauded by capital. Without a notion of individual property, commodities, as we know them, could not exist. There would only be users without owners.

Today, copyright law, the most stringent of property laws, ensures market space for commodities. Due to the rise of Internet file-sharing, however, it is in a state of crisis. In our present day of ever-monopolizing, wealthy media conglomerates, file-sharing has crippled the music industry, which, to say the least, is an amazing feat. Those who did not wish to spend $20 on a CD could easily download that very same cultural product from Napster, Kazaa!, Soulseek, eMule, or BitTorrent (which by itself takes up over a third of Internet bandwidth today)[22] and, what was even worse, they could then share it with others as well. Reproduction was occurring without the oversight of capital's own system, and thus without the production of profit for the CEOs of the media conglomerates.

Downloading poses interesting questions for us; it forces us to rethink the commodity as well as the labor that goes into its reproduction. The technology of the mp3, short for MPEG Audio Layer 3, a digital encoding system for audio, serves to illustrate our entrenchment in a new virtual culture highly dependent on new media technologies as well as the integration and interoperability of technomedia beyond individual "home" systems. Technomedia demand that technologies be able to read each other's formats. This is the very point made by McLuhan regarding new and old media. All new technologies fold into one another. The brilliance of the mp3 (or scariness, for music industry executives) is its almost universal readability. The mp3 is easily the most widely playable format in the world. In the United States, the portable audio industry has exceeded the home and car audio industries combined.[23] On top of the enormous rise in mp3 player sales, mp3 files can be played by an increasingly wide variety of electronic devices. Besides computers and mp3 players, mobile phones, portable video game devices, DVD players, and CD players can play mp3s. Today, cars come equipped to allow mp3 players to plug into their stereo systems, indicating the interoperable nature of technomedia. The mp3 is not the only such format. Digital video is now compressed into a file format known as AVI (Audio Video Interleave) or QuickTime for easy playback on devices that we can no longer call "mp3 players." The leading portable media players, including the iPod, play back video and music and also hold digital photos. They are, in true technomedia form, multiple media players. Moreover, mp3 and AVI files are difficult to trace, and easily bootlegged, because of their digital format. Reproduction in the virtual era has become simply holding down CTRL-C on the keyboard.

Network and Informational Politics

In their widely read *Empire,* Michael Hardt and Antonio Negri identify the Internet as a forum for contentious politics precisely due to its "deterritorialized structure." The Internet is inherently democratic because of its *network* structure. The network is different from previous hierarchical structures because it does not privilege a central node of control. In contrast, it is "an indeterminate and potentially unlimited number of interconnected nodes."[24] Each node is able to operate on its own in an autonomous fashion; the network does not fall apart even when all of its parts are not functioning. Therefore, no operation within the network is centrally controlled. This informational structure decentralizes production. "Informational production and the network structure of organization," Hardt and Negri suggest, "make productive cooperation and efficiency no longer dependent . . . on proximity and centralization."[25]

However, they uncharacteristically embed their notion of the decentralization of production on the Internet within a hope for the reconstruction of "the commons," which harkens back to my earlier discussion of Habermas's Enlightenment-oriented "rational public sphere." By arguing that production has been "informatized," Hardt and Negri fall into the same trap as Lazzarato by unabashed strict adherence to Marx's vision of historical progress.[26] They see the Internet-as-commons within a linear narrative of development leading to the emancipation of the multitude.[27] They even quote Rousseau to make their point:

> . . . the first person who wanted a piece of nature as his or her own exclusive possession and transformed it into

Michael Hardt (1960–) and **Antonio Negri** (1933–) are the co-authors of three books—*The Labor of Dionysus* (1994), the best-selling *Empire* (2000), and its sequel *Multitude* (2004). While working toward his doctorate at the University of Washington with Fredric Jameson, Hardt visited Negri in Paris and later went on to write his dissertation under his supervision. Negri had been in Paris as a political exile from his native Italy, where he had been accused of trumped-up charges of antistate activity. Before he had been arrested and forced to flee, he had been professor of political science at the University of Padua. In Italy, Negri, with Mario Tronti, had been the founder of the *operaismo* movement, which is seen as the precursor to Autonomia Marxism. His commitments to *operaismo* and Autonomia are clear in the works he co-authored with Michael Hardt as well as in the works of subsequent thinkers affiliated with them—namely Paolo Virno and Maurizio Lazzarato. With *Empire*, Hardt and Negri achieved the rare feat of finding a large audience for an academic work. It managed to blend radical politics, high-theory, and, of course, a sweeping critique of globalization, with original insight.

Multitude: The *multitude* is a concept found in the work of the philosopher Baruch Spinoza, which Hardt and Negri theorize as a break from the dominant form of subjectivity found in traditional Marxism—the proletariat as defined by class position as determined by the exploitative labor process. Hardt and Negri use multitude to contest the formulation of "the people," rooted in a notion of sovereign transcendence, which they find to be an inherent limitation in Marx's concept of the revolutionary "working class." In fact, the "multitude," as a revolutionary subjectivity, draws as much from Deleuze as it does from Spinoza in that Hardt and Negri contend that it is immanent, or virtual, requiring it to be what Negri calls, "*unrepresentable singularities* rather than *individual* proprietors."[28]

> the transcendent form of private property was the one who invented evil. Good, on the contrary, is what is common.[29]

They emphasize the potential for a struggle against Empire based on the postmechanical reproductive capability of technomedia, what they describe as "decentralization" and "deterritorialization." But in doing so, they maintain a distinction between "original" and "copy" that does not take into consideration the changes in the informational dynamics to which Terranova alludes. The new informational milieu pushes us beyond the notion of information as simply the content of communication, to which Hardt and Negri still hold very tightly. By maintaining that commodities (be they services or material goods) must be shared as they had been in the feudal commons, Hardt and Negri are separating the good or commodity in postmodern production (audio or video files, for example) from their information dynamics—their "file properties" so to speak.

The increasing importance of such properties cannot be overemphasized when they determine a commodity's mode of reproduction. For example, I could obtain Korean director Park Chan-wook's film *Old Boy* (2003) through several routes: (1) I could rent it, either through my local video rental store or through Netflix, a home delivery service. (2) I could buy it, through Amazon or any other retailer. (3) I could download it using BitTorrent. Though I am obtaining, technically speaking, the same product—the film—I am also getting along with it two different informational dynamics. If I choose either of the first two routes, I receive a DVD. With a DVD, my ability to reproduce the film illegally or, in other words, to pirate it, requires a certain type of technological solution. As are most DVDs today, this particular film would most likely be copyright-protected by a digital-encoded system. I would have to make use of "ripping" software—software that extracts audio or video data to a hard disk. My third route is more conducive to pirating. If I have obtained the film in digital format, someone else has already done the ripping for me. The file–properties, or what Terranova calls "informational dynamics"[30] more generally, are gaining greater influence in the process of *reproduction* particularly.

File-sharing can occur on a mass scale because of the extant interoperability of mp3s between different electronic devices such as computers, mobile phones, mp3 players, and flash drives. Multiple devices can read, play, and copy mp3s. This type of digital reproduction occurs within truncated space and time. The simple action of CTRL-C (keyboard shortcut command for copy) and CTRL-V (shortcut for paste) is quite different from, let's say, the days of dubbing tapes on a tape deck. Not only was dubbing slow (even with the integration of high-speed dubbing), the sound quality would deteriorate with each copy. On the contrary, reproduction of mp3s occurs without a compromise in sound quality (bit rate). The emphasis in the virtual era has shifted from production to reproduction. Though Poster and Lazzarato shrewdly note the changes in production, Terranova recognizes a modulation in the logics of capital itself in which reproduction gains significance. And this new form of reproduction calls forth a correlative politics by which many can consume without making an attempt to participate in the means of production. With the capacity to reproduce music, movies, and other commodities at rates of 400–480 Mbits/second (FireWire and USB2.0), there is potential for the consumer to also act as *reproducer* and for the value of a given commodity to be driven to nearly zero.

Hip-Hop's Culturo-Informational Politics: Sampling

The history of sampling in Western music dates at least back to jazz music, in which "standards" or show tunes from Broadway musicals would be played and interpreted by black musicians. Instead of sticking to the lyrics and sheet music verbatim, jazz musicians would play the melody with their own inflections along with their own improvisations. Similarly, hip-hop DJs originally used looped "breaks" from disco songs in order to make beats, the musical backbone on which emcees, or rappers, would rhyme. This technique of looping is a crude form of the nuanced sampling that now occurs in and out of the genre of hip-hop. The DJ, utilizing a two-channel mixer with a cross fader and two turntables, was able to loop a "break." In hip-hop, a "break" refers to a section of a song that is purely instrumental, with no vocals. The early DJs took two copies of the same record to loop these sections to provide an empty musical canvas for rappers. Many contemporary producers have utilized this old school method of sampling as well. Producers Sean "P. Diddy" Combs and Chucky Thompson looped elements of "In Between the Sheets" by the Isley Brothers, a hit R&B record from the 1970s, for one of the most recognizable songs of 1990s-era hip-hop, "Big Poppa" by Notorious B.I.G.

Sampling, as hip-hop critic Nelson George notes, cemented hip-hop music as "postmodern art in that it *shamelessly* raids older forms of pop culture"[31] by recontextualizing sounds. The Isley Brothers could not have foreseen a revival in their careers during the 1990s due to a slew of samples taken from hits they recorded in the 1970s. It also recontextualized or, as Siva Vaidhyanathan describes, riffing on Public Enemy, "bumrushed" copyright law.[32] Vaidhyanathan argues in *Copyrights and Copywrongs* that sampling reveals the inadequacy of copyright law in the midst of new communications technologies, techniques, and *aesthetics*.[33] George and Vaidhyanathan agree that digital sampling technology, and its aesthetic of "mishmash mosaic," is rooted in the very informational dynamics that Terranova sought to identify. To put it differently, digital technologies brought with them the aesthetics of sampling that challenged not only the dominant song-structure of Anglo-American pop music (exemplified by the music of the Beatles), but the project of modernity itself, by reimagining aesthetic possibilities through technology. Hip-hop, through sampling, became a force of resistance that was first and foremost cultural, and by necessity social and political.

Hip-hop's challenge to copyright law—digital sampling—started with the early digitalization technologies. Today, sampling in hip-hop, as well as throughout other genres of music, is dominated by computer "soft-samplers" (software samplers used in digital recording/sequencing workstations on computers). But in the late 1970s and early 1980s, only hardware samplers were available. The earliest samplers were produced by Fairlight Computer Musical Instruments and E-mu, two synthesizer companies.[34] Later on, the samplers of choice for hip-hop producers were the Akai MPC-2000, Ensoniq ASR, and the E-mu SP-1200. In the process of sampling, computers convert sounds into binary units readable by computers. This conversion allows the "producer"—the popular term for composer of the music in hip-hop—to tweak and loop the sounds as he or she pleases. The *bricolage* method of hip-hop music contested "the very definitions of 'work', 'author' and 'original'," the fundamental "basics of copyright law,"[35] as Vaidhyanathan writes. These categories were also fundamental to modern capitalism.

The categories listed by Vaidhyanathan—"work," "author," and "original"—are all objects of scrutiny for Roland Barthes in "The Death of the Author." In this essay, Barthes makes a sharp critique of what Foucault called the "author-function" as well as the notion of autobiography in text or a work of literature.[36] He notes that "the image of literature to be found in ordinary culture is *tyrannically centred on the author*, his person, his life, his passions."[37] This notion of authorship in literature explains also the obsession with copyright law, which preserves the concept of one proper and originary author for one work. As Barthes points to Mallarmé, Valery, and Proust as those who challenged the dominance of the Author, I point to Public Enemy's production team the Bomb Squad and contemporary glitch-hop musician Prefuse 73 as those who undermine the restriction of cultural products by sampling.[38] Sampling pulls down forcibly the holiness of the work of music from its ivory tower, where it remained through the era of mechanical production in spite of observations made by Walter Benjamin. Though *aura* may have decayed, the concept of authorship nevertheless remained of value despite the massive transformations in commodity-reproduction techniques prevalent in the early twentieth century (Fordism and other related automated technologies).

The 1980s saw a host of high-profile lawsuits involving hip-hop groups over the issue of sampling. One such case was that of the Turtles, a rock group, who sued De La Soul, a rap group from Long Island, New York, for using elements of their song "You Showed Me." Though settled out of court for a large sum of money, the case represented a defeat that would change the way hip-hop music was produced. Samples needed to be cleared from then on by the "authors" of the "original" work. Clearance issues have long since plagued hip-hop music and have occasionally barred songs from being included on artists' albums due to a lack of permission. Phife Dawg, from rap group A Tribe Called Quest, expresses this frustration, lamenting in "8 Million Stories": "Stressed out more than anyone could ever be / Forever tryin' to clear the samples for my new LP."[39]

However, sample clearance difficulties brought rise to another type of music production that avoided clearing of samples altogether. While based on loops early on, hip-hop beats began to take a different form as DJ culture, or turntablism,[40] developed within hip-hop. Breaking away from looping breaks, DJs began to use a method perfected by the Bomb Squad. Using the latest technology for music production, the Bomb Squad used several samples to create a single song by sampling different elements from many different songs, not just one. In any given song by DJ Shadow—probably the most well-known musician in this vein today—there could be up to ten different songs from which he had sampled different musical elements. ("Digging," or looking for the songs that producers sampled, is now a subculture of its own.) A four-bar drum loop here, the tail end of a saxophone solo there, strings from an opera, ambient noises of the street. Every sound became a potential sample. In addition, the continuous innovations in digital sampling technology allowed samplers to be, in music production terms, polyphonic. They could stretch samples to vary their pitch and speed, much like synthesizers did, allowing a single noise to be played through the entire tonal range of the keyboard (of the ivory and ebony variety, not QWERTY). This type of innovation allowed producers to mask or tweak samples to the point of unrecognizability, enabling them to take elements from multiple sources and reproduce them without consent.

Decentering, or Modulations in Technomedia

Sampling and downloading act as new forms of an exemplary blend of a cultural politics. In this digital moment, within the ubiquity of technomedia—where information has not only carved out the environment, but is the environment itself—culture is increasingly influential in shaping the social. At the intersections of culture and technology, there is an opening for a politics against science, reason, and law—the bedrocks of modernity. Sampling and downloading reclaim the mass production logic of capitalism today by undercutting the profitability of increasingly expensive cultural products. Beyond the interpretation of this type of politics as the underclass appropriating reproduction under the radar of media conglomerates, and beyond the classical Marxist dream of reclaiming the means of production, there is ultimately the virtual technological "bumrushing" of the means of cultural and informational reproduction by the file-sharers, which modulates—or decenters in Derridean terms—away from domination through culture.

In a short but telling article written near the end of his life, Deleuze states that the disciplinary societies of the eighteenth and nineteenth centuries that Foucault described are supplanted in the twentieth century by "societies of control."[41] The logic of control—modulations of continuously changing spaces and times—differs from the logic of discipline, which is more akin to molds (e.g., family, school, prison, hospital). The work of a politics in this era of spatio-temporal modulations is not to appeal to ideals of representation and inclusion into spaces, the very institutions of the controlling system, because, in effect, there is no choice but to be included; we are already plugged in.

In such a society, to celebrate "the copy" for its "unoriginality" and instability is the source of a politics. Judith Butler's masterful critique of gender "identity" as drag, or imitation, is of particular relevance as a concluding thought

> **Judith Butler** (1956–) is Maxine Elliot Professor in the Department of Rhetoric at the University of California, Berkeley. Perhaps the most prolific queer theorist in the United States, she is the author of *Gender Trouble* (1990), the seminal work in the field. She has also written on hate-speech, mourning, and subjectivity. She often utilizes poststructuralist and psychoanalytic theories as starting points for the development of her own. Her theory of gender performativity is discussed below.

about the politics of modulation. Though Butler does not speak directly about the Internet and other forms of technomedia, her discussion of sexual politics provides a useful starting point.[42] Butler writes:

> Compulsory heterosexuality sets itself up as the original, the true, the authentic; the norm that determines the real implies that "being" lesbian is always a kind of miming, a vain effort to participate in the phantasmatic plenitude of naturalized heterosexuality.[43]

Butler's use of the word "drag" is indispensable, in spite of its popular usage as an adjective for "queens," men who take on feminine personas, which of course includes feminine attire as well. Butler, following Esther Newton, uproots the meaning of "drag," arguing that it is "not the putting on of a gender that belongs properly to some other group."[44] Therefore, it is not an appropriation of one proper sex's "cultural property" by another. It is, as Butler notes, "the mundane way in which genders are appropriated, theatricalized, worn and done," and concludes, "that all gendering is a kind of impersonation and approximation."[45] She posits that gender is not the performance of a given essence of sexuality from within the individual but an imitation that lacks an original. The notion of the original is, as Butler shows, an *effect* of the imitations that "posture as grounds, origins, the normative measure of the real."[46] Hence, homosexuality-as-imitation is a constructive necessity for heterosexuality, in itself "a *panicked* imitation of its own idealization."[47]

This is surely the case with what Vaidhyanathan identifies with the rise of copyright law on the Internet. The response to file-sharing by corporate capital has been to reinscribe downloading from an illegal activity to a legal, paid-for service like the Apple iTunes music store. The iTunes music store is an example of the panicked imitations of which Butler speaks. Recently, as the record industry saw its sales dip to record lows, it reacted by brokering similar deals with online music stores and even pursuing legal action. The RIAA (Recording Industry Association of America) and the MPAA, its Hollywood equivalent, have directed enormous efforts toward keeping copyrighted material offline altogether. Recent crackdowns on MySpace and YouTube are clear evidence of the fear of copyrighted media's *aura* being undone. Undoubtedly, it is already too late.

Though they disagree on whether to call it late modern, liquid modern, reflexive modern, postmodern, postideological, postindustrial, and so on, many authors agree that the present era lacks thc potential for a radical politics. Some, like Ernesto Laclau, Chantal Mouffe, Antonio Negri, Michael Hardt, and Jacques Ranciere have deemed our era postpolitical. In the wake of global finance capital, neoliberalism, and the IT revolution, it seems that the world has become too complicated to suit the older forms of political alliances. In the last analysis, Marx's famous words claiming that "all things solid melt into air" may prove to be the case for politics too. It is clear that "the Left" and "the Right" are no longer viable descriptors of sound political positions. In American electoral politics, the rise of the Washington Consensus points to such a diagnosis of the times. In the UK, the Third Way policies of the (New) Labour Party play much a similar role.

Media technologies have been portrayed either as the root of this predicament or the solution. Some see them as taking on the democratic movements of the 1960s. Because of its anti-hierarchical structure, the Internet is afforded potential for left politics. Others see it as a fragmentary force that distracts from more than facilitates political activity. These arguments are two sides of the same coin. Each assumes that the Internet and other technomedia will bend to the extant model of politics. The guiding wisdom throughout this book is that technologies are not simply tools to be mastered by humans for human ends. As did the early DJs of hip-hop, we must derive politics *from* the informational features of media technologies to achieve desired effects.

The early block parties, where hip-hop culture was bred, kept on into the night by stealing electricity from the street lamps on the sidewalks of New York City to power the turntables and speakers. In that truly crafty spirit of making art from life, politics in technomedia must not shy away from the technologies that have already changed the world in profound ways, but accept them in a way that resembles what Lévinas refers to as the ethics of the Encounter, and which Derrida interprets as the radical gesture of hospitality, so as not to impart the vulgar arrogance of eras past.

Notes

1. Fredric Jameson, "Postmodernism and Consumer Society," in *The Anti-Aesthetic: Essays on Postmodern Culture*, ed. Hal Foster, 1st. ed. (Port Townsend, Wash.: Bay Press, 1983).

2. Benedict Anderson draws a connection between the advent of the printing press and the emergence of nationalism (and along with it a "comradeship" based on culture). See his *Imagined Communities: Reflections on the Origin and Spread of Nationalism* (London: Verso, 1983), 5–7.

3. Charles Lemert, *Postmodernism Is Not What You Think* (Oxford; Cambridge, Mass.: Blackwell, 1997), 28.

4. I am offering a reading of Jameson's periodization of postmodernism to question and discuss the centralizing tendency of his critique of postmodernism and, in particular, his argument that postmodernism is a cultural dominant.

5. Fredric Jameson, "Postmodernism or the Cultural Logic of Late-Capitalism," in *Media and Cultural Studies: KeyWorks*, ed. Meenakshi Gigi Durham and Douglas Kellner, rev. ed. of *KeyWorks in Cultural Studies* (Malden, Mass.: Blackwell, 2006), 486.

6. Tiziana Terranova, "Communication Beyond Meaning: On the Cultural Politics of Information," *Social Text* 22, no. 3 (2004): 52.

7. Terranova, "Beyond Meaning," 53.

8. Terranova's discussion does not mention debates within Marxist scholarship. Her critique tends to homogenize Marxist thought as uncritical of Marx. However, historical materialism, for one, has been up for discussion in Marxist thought for some time. See, for example, Stanley Aronowitz, *The Crisis in Historical Materialism: Class, Politics, and Culture in Marxist Theory*, 2nd ed. (Minneapolis: University of Minnesota Press, 1990).

9. Terranova, "Beyond Meaning," 54.

10. Karl Marx, *A Contribution to the Critique of Political Economy* (New York: International Publishers, 1970), 20–21. Emphasis mine. Thanks to Jonathan Cutler for bringing this passage to my attention.

11. Mark Poster, *The Mode of Information: Poststructuralism and Social Context* (Cambridge, UK: Polity Press, 1990), 5.

12. Poster, *Mode of Information*, 7.

13. Poster, *Mode of Information*, 8.

14. Maurizio Lazzarato, "Immaterial Labor," in *Radical Thought in Italy: A Potential Politics*, ed. Virno Paolo and Michael Hardt (Minneapolis: University of Minnesota Press, 1996), 133–150, at www.generation-online.org/c/fcimmateriallabour3.htm.

15. Michael Hardt, "Affective Labor," *boundary 2*, 26, no. 2 (1999): 89–100.

16. Hardt, "Affective Labor," 89–100.
17. Hardt, "Affective Labor," 89–100.
18. Hardt, "Affective Labor," 89–100.
19. Lazzarato, "Immaterial Labor."
20. Lazzarato, "Immaterial Labor."
21. Lazzarato, "Immaterial Labor."
22. Spencer Kelly, "BitTorrent Battles over Bandwidth," *BBC News Online*, April 13, 2006, at news.bbc.co.uk/2/hi/programmes/click_online/4905660.stm.
23. Joseph Palenchar, "Portables Poised to Dominate Audio Industry, Statistics Show," *Twice*, November 7, 2005 at www.twice.com/article/CA6281863.html.
24. Michael Hardt and Antonio Negri, *Empire* (Cambridge, Mass.; London: Harvard University Press, 2000), 299.
25. Hardt and Negri, *Empire*, 295.
26. For an insightful critique of Hardt and Negri's *Empire* and *Multitude*, see Slavoj Zizek's "*Objet a* as Inherent Limit to Capitalism: on Michael Hardt and Antonio Negri," *Lacan.com* (Fall 2005), at www.lacan.com/zizmultitude.htm. Here, Zizek argues that the limitation in their argument coincides with Marx's scheme of historical progress, which necessitates the industrialization of feudal society in order to set up the coming revolution, which would of course lead, thought Marx, to a utopian communism.
27. Hardt and Negri, *Empire*, 301.
28. Antonio Negri, "Approximations: Towards an Ontological Definition of Multitude," *Multitudes* no. 9 (2003): 36–48.
29. As quoted in Hardt and Negri, *Empire*, 303.
30. Tiziana Terranova, *Network Culture: Politics for the Information Age* (London: Pluto Press, 2004), 6.
31. Nelson George, *Hip Hop America* (New York: Viking, 1998), viii.
32. Siva Vaidhyanathan, *Copyrights and Copywrongs: The Rise of Intellectual Property and How It Threatens Creativity* (New York: New York University Press, 2001), 132.
33. Vaidhyanathan, *Copywrongs*, 133. Emphasis mine.

34. George, *Hip Hop*, 92.
35. Vaidhyanathan, *Copywrongs*, 139.
36. I use "author-function" here to draw a connection between Foucault and Barthes in recognizing the primary problematic of "text." They, however, did have a debate concerning this very point and did not wholly agree. For a Foucauldian analysis of text, or what he labels "discourse," see "What is an Author?" in *Language, Counter-Memory, Practice: Selected Essays and Interviews* (Ithaca, N.Y.: Cornell University Press, 1977), 113–138.
37. Roland Barthes, "The Death of an Author," in *Image, Music, Text*, Roland Barthes and Stephen Heath (London: Fontana, 1977), 142–148.
38. The Bomb Squad is the name of a hip-hop production team that worked primarily with rap group Public Enemy. Their production style is noted for the layering of several samples on a single track, creating a thick and atonal style.
39. A Tribe Called Quest, "8 Million Stories," *Midnight Marauders* (Jive Records, 1993).
40. Turntablism is a term coined by DJ Supreme in 1994 to distinguish it from DJing—simply playing records. It consists of manipulating sounds using vinyl records and a turntable as a musical instrument. Though there are too many albums to list as good examples of turntablism, a good starter is Mix Master Mike's *Anti-Theft Device* (Asphodel, 1998).
41. Gilles Deleuze, "Postscript on the Societies of Control," *October*, no. 59 (1992): 3–7.
42. Judith Butler, "Imitation and Gender Insubordination," in *Social Theory: The Multicultural and Classical Readings*, ed. Charles Lemert, 3rd ed. (Boulder, Colo.: Westview Press, 2004), 562.
43. Butler, "Imitation," 562.
44. Butler, "Imitation," 563.
45. Butler, "Imitation," 563.
46. Butler, "Imitation," 563.
47. Butler, "Imitation," 564.

Selected Bibliography

Adorno, Theodor. "Culture Industry Reconsidered." *New German Critique* 6 (1975): 12–19.

Anderson, Benedict. *Imagined Communities: Reflections on the Origin and Spread of Nationalism*. London: Verso, 1983.

Aronowitz, Stanley. *Science as Power: Discourse and Ideology in Modern Society*. Minneapolis: University of Minnesota Press, 1988.

———, and Jonathan Cutler. *Post-Work: The Wages of Cybernation*. New York; London: Routledge, 1998.

———, and William DiFazio. *The Jobless Future: Sci-Tech and the Dogma of Work*. Minneapolis: University of Minnesota Press, 1994.

Barnet, Belinda. "Infomobility and Technics: Some Travel Notes." *CTHEORY* October 27, 2005, www.ctheory.net/articles.aspx?id=492.

Barthes, Roland. *Mythologies*. Paris: Editions du Seuil, 1957.

———, and Stephen Heath. *Image, Music, Text*. London: Fontana, 1977.

Bateson, Gregory. *Mind and Nature: A Necessary Unity*. 1st ed. New York: Dutton, 1979.

Baudrillard, Jean, and Mark Poster. *Jean Baudrillard: Selected Writings*. Stanford, Calif.: Stanford University Press, 1988.

Bauman, Zygmunt. *Liquid Modernity*. Cambridge, UK; Malden, Mass.: Polity Press; Blackwell, 2000.

———. "Time and Space Reunited." *Time and Society* 9, no. 2/3 (2000): 171–185.

Benjamin, Walter. *Illuminations*. New York: Schocken Books, 1968.

Braudel, Fernand. *On History*. Chicago: University of Chicago Press, 1980.

Clough, Patricia Ticineto. *Autoaffection: Unconscious Thought in the Age of Teletechnology*. Minneapolis: University of Minnesota Press, 2000.

———. *Feminist Thought: Desire, Power, and Academic Discourse*. Oxford: Blackwell, 1994.

De Landa, Manuel. *Intensive Science and Virtual Philosophy*. London; New York: Continuum, 2002.

Deleuze, Gilles. *The Fold: Leibniz and the Baroque*. Minneapolis: University of Minnesota Press, 1993.

———. "Postscript on the Societies of Control." *October*, no. 59 (1992): 3–7.

Derrida, Jacques, and Peggy Kamuf. *A Derrida Reader: Between the Blinds*. New York: Columbia University Press, 1991.

Dienst, Richard. *Still Life in Real Time: Theory after Television, Post-Contemporary Interventions*. Durham, N.C.: Duke University Press, 1994.

Durham, Meenakshi Gigi, and Douglas Kellner. *Media and Cultural Studies: KeyWorks*. Rev. ed. of *KeyWorks in Cultural Studies*. Malden, Mass.: Blackwell, 2006.

Eco, Umberto. *A Theory of Semiotics*. Bloomington: Indiana University Press, 1976.

Einstein, Albert. *Relativity: The Special and the General Theory*. London: Methuen, 1920.

Foster, Hal. *The Anti-Aesthetic: Essays on Postmodern Culture*. 1st. ed. Port Townsend, Wash.: Bay Press, 1983.

Foucault, Michel. *Discipline and Punish: The Birth of the Prison.* 2nd Vintage Books ed. New York: Vintage Books, 1995.

———. *Language, Counter-Memory, Practice: Selected Essays and Interviews.* Ithaca, N.Y.: Cornell University Press, 1977.

———. *The Order of Things: An Archaeology of the Human Sciences, World of Man.* London: Tavistock, 1970.

Freud, Sigmund. *Civilization and Its Discontents.* New York: W. W. Norton, 1961.

Friedman, Thomas L. *The Lexus and the Olive Tree.* 1st ed. New York: Farrar, Straus and Giroux, 1999.

George, Nelson. *Hip Hop America.* New York: Viking, 1998.

Giddens, Anthony. *The Constitution of Society: Introduction of the Theory of Structuration.* Berkeley: University of California Press, 1984.

Gouldner, Alvin W. *The Dialectic of Ideology and Technology: The Origins, Grammar and Future of Ideology.* London: Macmillan, 1976.

Grosz, E. A. *Becomings: Explorations in Time, Memory, and Futures.* Ithaca, N.Y.: Cornell University Press, 1999.

Guattari, Félix, and Gary Genosko. *The Guattari Reader.* Blackwell Readers. Oxford; Cambridge, Mass.: Blackwell, 1996.

Habermas, Jurgen. "The Public Sphere: An Encyclopedia Article." In *Media and Cultural Studies: KeyWorks*, edited by Meenakshi Gigi Durham and Douglas Kellner, 102–108. Rev. ed. of *KeyWorks in Cultural Studies.* Malden, Mass.: Blackwell, 2006.

Hacking, Ian. *Representing and Intervening: Introductory Topics in the Philosophy of Natural Science.* Cambridge: Cambridge University Press, 1983.

Haraway, Donna. "A Manifesto for Cyborgs: Science, Technology and Socialist Feminism in the 1980s." *Socialist Review* no. 80 (1985): 65–108.

Hardt, Michael, and Antonio Negri. *Empire.* Cambridge, Mass.; London: Harvard University Press, 2000.

———. *Multitude: War and Democracy in the Age of Empire.* New York: Penguin, 2004.

Hayles, N. Katherine. *How We Became Posthuman: Virtual Bodies in Cybernetics, Literature, and Informatics*. Chicago: University of Chicago Press, 1999.

Heidegger, Martin. *The Concept of Time*. Oxford; Cambridge, Mass.: Blackwell, 1992.

———. *The Question Concerning Technology, and Other Essays*. 1st ed. New York: Harper & Row, 1977.

Jay, Martin. "The Rise of Hermeneutics and the Crisis of Ocularcentrism." In "The Rhetoric of Interpretation and the Interpretation of Rhetoric," *Poetics Today* 9, no. 2, (1988): 307–326.

Jhally, Sut. *The Codes of Advertising: Fetishism and the Political Economy of Meaning in the Consumer Society*. London; New York: Routledge, 1990.

Kaplan, David M. *Readings in the Philosophy of Technology*. Lanham, Md.: Rowman & Littlefield, 2004.

Kittler, Friedrich. "The History of Communication Media." *CTHEORY* July 30, 1996, www.ctheory.net/articles.aspx?id=45.

Kristeva, Julia, and Kelly Oliver. *The Portable Kristeva, European Perspectives*. New York: Columbia University Press, 1997.

Lacan, Jacques. *The Four Fundamental Concepts of Psycho-Analysis*. New York; London: W. W. Norton, 1998.

Lazzarato, Maurizio. "Immaterial Labor." In *Radical Thought in Italy: A Potential Politics*, edited by Paolo Virno and Michael Hardt. Minneapolis: University of Minnesota Press, 1996.

Lefebvre, Henri. *The Production of Space*. Oxford; Cambridge, Mass.: Blackwell, 1991.

Lemert, Charles. *Postmodernism Is Not What You Think: Twentieth-Century Social Theory*. Oxford; Cambridge, Mass.: Blackwell, 1997.

———. *Social Theory: The Multicultural and Classic Readings*. 3rd ed. Boulder, Colo.: Westview, 2004.

———, and Anthony Elliott. *Deadly Worlds: The Emotional Costs of Globalization*. Lanham, Md.: Rowman & Littlefield, 2006.

Lévinas, Emmanuel. *Time and the Other and Additional Essays*. Pittsburgh, Penn.: Duquesne University Press, 1987.

Lutticken, Sven. "Suspense and . . . Surprise." *New Left Review* no. 40 (2006): 95–109.

Marx, Karl. *A Contribution to the Critique of Political Economy*. New York: International Publishers, 1970.

McLuhan, Marshall. *The Gutenberg Galaxy: The Making of Typographic Man*. Toronto; London: University of Toronto Press, 1962.

———. *Understanding Media: The Extensions of Man*. New York: McGraw-Hill, 1964.

Misa, Thomas J., Philip Brey, and Andrew Feenberg. *Modernity and Technology*. Cambridge, Mass.: MIT Press, 2003.

Munster, Anna, and Geert Lovink. "Theses on Distributed Aesthetics. Or, What a Network Is Not." *fibreculture* no. 7 (2005), journal.fibreculture.org/issue7/issue7_munster_lovink.html.

Ong, Walter J. *Orality and Literacy: The Technologizing of the Word*. London: Routledge, 1993.

Poster, Mark. *The Mode of Information: Poststructuralism and Social Context*. Cambridge, UK: Polity Press, 1990.

———. *What's the Matter with the Internet?* Minneapolis: University of Minnesota Press, 2001.

Postman, Neil. *Technopoly: The Surrender of Culture to Technology*. 1st ed. New York: Knopf, 1992.

Rabinbach, Anson. *The Human Motor: Energy, Fatigue, and the Origins of Modernity*. Berkeley: University of California Press, 1992.

Ronell, Avital. *The Telephone Book: Technology—Schizophrenia—Electric Speech*. Lincoln: University of Nebraska Press, 1989.

Schuerewegen, Franc. "A Telephone Conversation: Fragments." *Diacritics* 24, no. 4 (1994): 30–40.

Sontag, Susan, and Geoffrey Movius. "An Interview with Susan Sontag." *The Boston Review* June 1975.

Spooky, DJ. "Loops of Perception: Sampling, Memory and the Semantic Web." *Horizon Zero* Issue 8 (2003).

Terranova, Tiziana. "Communication Beyond Meaning: On the Cultural Politics of Information." *Social Text* 22, no. 3 (2004): 51–73.

——. *Network Culture: Politics for the Information Age*. London: Pluto Press, 2004.
Thompson, John B. *The Media and Modernity: A Social Theory of the Media*. Cambridge, UK: Polity Press, 1995.
Vaidhyanathan, Siva. *Copyrights and Copywrongs: The Rise of Intellectual Property and How It Threatens Creativity*. New York: New York University Press, 2001.
Virilio, Paul. *The Information Bomb*. London; New York: Verso, 2000.
——, and James Der Derian. *The Virilio Reader*. Blackwell Readers. Oxford; Malden, Mass.: Blackwell, 1998.
Virno, Paolo, and Michael Hardt. *Radical Thought in Italy: A Potential Politics*. Minneapolis: University of Minnesota Press, 1996.
Wallerstein, Immanuel. *The Uncertainties of Knowledge, Politics, History, and Social Change*. Philadelphia: Temple University Press, 2004.
Weber, Samuel, and Alan Cholodenko. *Mass Mediauras: Form, Technics, Media*. Stanford, Calif.: Stanford University Press, 1996.
White, Michele. *The Body and the Screen: Theories of Internet Spectatorship*. Cambridge, Mass.: MIT Press, 2006.
Wiener, Norbert. *The Human Use of Human Beings: Cybernetics and Society*. New York: Avon Books, 1967.
Zizek, Slavoj. *The Ticklish Subject: An Essay in Political Ontology*. New York: Verso, 1999.
——. "The Interpassive Subject." *Traverses* 1998.
Zweig, Ellen. "The Lurker: Outline for a Murder Mystery." In *Being Online: Net Subjectivity*, edited by Alan Sondheim. New York: Lusitania, 1996.

Index